Ontological Analyses in Science, Technology and Informatics

Edited by Andino Maseleno and Marini Othman

Published in London, United Kingdom

IntechOpen

Supporting open minds since 2005

Ontological Analyses in Science, Technology and Informatics
http://dx.doi.org/10.5772/intechopen.77594
Edited by Andino Maseleno and Marini Othman

Contributors
Bohdan Hejna, Martin Žáček, Alena Lukasová, Marek Vajgl, Petr Raunigr, Sukumar Mandal, Friska Natalia, Dea Cheria, Santi Surya, JinTa Weng, Jing Qiu, Ying Gao

Notice
Statements and opinions expressed in the chapters are these of the individual contributors and not necessarily those of the editors or publisher. No responsibility is accepted for the accuracy of information contained in the published chapters. The publisher assumes no responsibility for any damage or injury to persons or property arising out of the use of any materials, instructions, methods or ideas contained in the book.

First published in London, United Kingdom, 2020 by IntechOpen
IntechOpen is the global imprint of INTECHOPEN LIMITED, registered in England and Wales, registration number: 11086078, 7th floor, 10 Lower Thames Street, London,
EC3R 6AF, United Kingdom
Printed in Croatia

British Library Cataloguing-in-Publication Data
A catalogue record for this book is available from the British Library

Additional hard and PDF copies can be obtained from orders@intechopen.com

Ontological Analyses in Science, Technology and Informatics
Edited by Andino Maseleno and Marini Othman
p. cm.
Print ISBN 978-1-78985-547-0
Online ISBN 978-1-78985-548-7
eBook (PDF) ISBN 978-1-83881-092-4

Meet the editors

Dr Andino Maseleno is a lecturer at STMIK Pringsewu, Lampung, Indonesia. He was a postdoctoral researcher at the Institute of Informatics and Computing Energy, Universiti Tenaga Nasional, Malaysia, from 2018 to 2019. He was a visiting fellow at the Centre for Lifelong Learning, Universiti Brunei Darussalam, Brunei, Darussalam, from July 2016 to March 2017. He received his BSc degree in Informatics Engineering from UPN Veteran Yogyakarta, Indonesia in 2005, his M.Eng. in Electrical Engineering from Gadjah Mada University, Indonesia in 2009, and his Ph.D. in Computer Science from Universiti Brunei Darussalam, Brunei Darussalam in 2015.

Professor Marini Othman is the director of the Institute of Informatics and Computing Energy, Universiti Tenaga Nasional, Malaysia. She received her BSc in Computer Science from Indiana State University, USA, in 1986, her MSc in Computer Science from Western Kentucky University, USA, in 1987, and her PhD in Industrial Computing from Universiti Kebangsaan Malaysia, Malaysia, in 2010.

Contents

Preface   XI

Chapter 1   1
Common Gnoseological Meaning of Gödel and Caratheodory Theorems
by Bohdan Hejna

Chapter 2   23
Knowledge Patterns within the Conception of Semantic Web
by Martin Žáček, Alena Lukasová, Marek Vajgl and Petr Raunigr

Chapter 3   35
Ontology Language XOL Used for Cross-Application Communication
by Jinta Weng, Jing Qiu and Ying Gao

Chapter 4   47
An Ontology-Based Approach to Diagnosis and Classification for an
Expert System in Health and Food
by Friska Natalia, Dea Cheria and Santi Surya

Chapter 5   67
Taxonomy and Ontology Management Tools: A General Explanation
by Sukumar Mandal

Preface

Ontology in the philosophical sense is concerned with the nature of being as well as with respective basic concepts. Aristotle called ontology the first philosophy. Despite its metaphysical nature, this first philosophy or ontology is a significant attempt to introduce a systematic approach to the process of thinking about the world surrounding us, about our conceptual thinking and about us ourselves.

Today's advanced science inherited the original ontology's efforts to systemize and conceptualize. In this respect the concept of ontology nowadays is a non-speculative methodology for studying reality objects and used tools, both of which are important for our orientation in the physical, technical, mental, and social worlds. As such, we deal with the process of studying, as well as with the outcomes of such study, of the objects observed or created by man and their respective concepts, relations between them, and relations between their systems in different fields of science.

Understood in this modern and precise way, this book applies the meaning of ontology in a search for semantic meanings of objects and concepts. It examines the relations between general or abstract terms and their meanings using a variety of logics such as automated indexation of databases. Therefore, the meaning of "to be" and "how to be" is exercised in complex or abstracted cases, for example, in number theory. The objective is to model, categorize, and conceptualize the knowledge available, and to design relevant schemes such as function-related diagrams or tree structures. Then, pending on the field of study, it is possible to discuss, introduce, and compare a variety of ontologies as the formal models representing our knowledge of the field or the problem. As such, it is possible to discuss ontology as it relates to experimental organization, software and systems engineering, artificial intelligence, the Semantic Web, and informatics and information sciences.

Through these kinds of ontologies we understand the models with standard structures of entities, classes, qualities, and relations. Such models are explicit, created using suitable language and formalized descriptions of the given systems of the objects or their respective concepts and relations. Then we speak about a model or a data model of the problems. The language used for these purposes can be formal (e.g., languages of physical or mathematical theories), semiformal, or informal, especially in the initial stages of studying the problems.

However, ontologies are not only the final formal and declarative representations, models, or data models of given problems but they are also the methodology and, consequently, the method and the process of creating these declarations or models. Ontology as a process creates, uses, and provides, as its output, a model or descriptive ontology. Such a model is a glossary of definitions of concepts corresponding to the objects, a thesaurus of definitions of the relations among the concepts and the respective objects. Thus, the ontology in this descriptive sense is both vocabulary and grammar that are used to keep and pass over the knowledge of the problems studied.

This book *Ontological Analyses in Science, Technology and Informatics* is the illustration of the modern and scientific application of ontology.

Chapter 1 deals with logic inference ontology used in the theory of proof based upon the language of physics; Chapter 2 examines knowledge-based pattern ontology in the Resource Description Framework (RDF) language in Semantic Web applications; Chapter 3 deals with implementation of the ontological XOL language in cross-application communication; Chapter 4 discusses ontological studies in the fields of diagnosis and expert systems in health and food; and Chapter 5 covers management ontology and taxonomy.

This book is designed for theorists as well as those persons dealing with designing, developing, managing, and decision-making in the field of ontology.

I would like to thank the publishing process managers and technical staff at IntechOpen for their commitment, friendly effort, and support throughout the production of this book. Finally, I express my thanks to all the contributing authors for their valuable chapters.

Ing. Bohdan Hejna, Ph.D., dr. h. c.
Department of Mathematics,
University of Chemistry and Technology,
Prague, Czech Republic

Chapter 1

Common Gnoseological Meaning of Gödel and Caratheodory Theorems

Bohdan Hejna

Abstract

We will demonstrate that the *I.* and *the II. Caratheodory theorems* and their common formulation as the *II. Law of Thermodynamics* are physically analogous with the *real* sense of the *Gödel's* wording of his *I.* and *II. incompleteness theorems*. By using physical terms of the *adiabatic changes* the Caratheodory theorems express the properties of the *Peano Arithmetic inferential process* (and even properties of any *deductive and recursively axiomatic* inference generally); as such, they set the physical and then logical limits of any real inference (of the sound, not paradoxical thinking), which can run only on a physical/thermodynamic basis having been compared with, or translated into the formulations of the Gödel's proof, they represent the first historical and *clear* statement of gnoseological limitations of the deductive and recursively axiomatic inference and sound thinking generally. We show that **semantically understood** and with the language of logic and meta-arithmetics, the full meaning of the **Gödel proof expresses the universal validity of the II. law of thermodynamics and that the Peano arithmetics is not self-referential and is consistent.**[1]

Keywords: arithmetic formula, thermodynamic state, adiabatic change, inference

1. Introduction

To show that the real/physical sense of the Gödel incompleteness theorems—that the very real sense of them—is the meta-arithmetic-logical analog of the Caratheodory's claims about the *adiabatic system* (that they are the analog of the sense of the *II. Law of Thermodynamics*), we compare the states in the *state space* of an *adiabatic thermodynamic system* with *arithmetic formulas* and the *Peano inference* is compared with the *adiabatic changes* within this state space. The *whole set of the states* now *not achievable adiabatically* represents the existence of the states on an adiabatic path, but this fact is not expressible adiabatically. This property of which is the

---

[1] The reader of the paper should be familiar with the Gödel proof's way and terminology; SMALL CAPITALS in the whole text mean the Gödel numbers and working with them. This chapter is based, mainly, on the [1–4]. This paper is the continuation of the lecture *Gödel Proof, Information Transfer and Thermodynamics* [4].

analog of the sense *of Gödel undecidable formula*. Nevertheless, any of these states, now not achievable adiabatically in the given state space (of the given adiabatic system), is achievable adiabatically *but* in the redefined and wider adiabatic system with its state space divided between adiabatic and not adiabatic parts again. These states (which are achievable only when the previous subsystem is part of the new actual system, both are consistent/adiabatic) represent arithmetic but not the Peano arithmetic formulas and also are bearing the property of their whole set. Also they can be axioms of the higher/superior inference including the previous one—the *general arithmetic inference* is further ruled by the same and repeated principle of widening the axiomatics and with same thermodynamic analogy using the redefined and widened new adiabatic system and its settings and with the same limitation by the impossibility to proof both the consistency of the given inferential system and, in our analogy, the adiabacity of its given adiabatic analog, by means of themselves. The consistency of the inferential system and adiabacity of its analog (and their abilities generally) are defined and proved by outer construction, outer limitations, and outer settings only (compare this our claim with the Gödel's claim for the Peano arithmetic inference " ... in the Peano arithmetic system exists ... ").

Caratheodory common formulation of the *II. P.T.*:

> *In the arbitrary vicinity of every state*
> *of the state space $\mathfrak{Q}_\varrho$ of the <u>adiabatic</u> system $\mathfrak{L}$*
> *exist states <u>not</u> reachable*
> *from the starting state adiabatically* ($[\mathrm{d}]\mathbf{Q}_{\mathbf{Ext}}=0$)
>
> *(or the states not reachable by the system at all).*

For the <u>consistency</u> of the **Peano arithmetic theory** $\mathcal{T}_{PA}$
the analog is expressed by:

Gödel incompletness theorems:

> *<u>For</u> the theory $\mathcal{T}_{PA}$ exists the true ("1") CLAIM*
> *that either this CLAIM and its NEGATION*
> *is <u>NOT</u> PROVABLE within the system $\mathcal{P}/\mathcal{T}_{PA}$.*

- *CLAIM* about the $\mathcal{T}_{PA}$ <u>consistency</u> especially -

> *The CLAIM saying that theory $\mathcal{T}_{PA}$ is <u>consistent</u>*
> *is <u>not</u> PROVABLE by its means* ($\mathcal{P}$) *- by itself.*

In our considerations, we use the states of the adiabatic system as the thermodynamic representation of the Peano arithmetically inferred formulas and the transition between the stats is then the thermodynamic model of the Peano arithmetic inference step, the consistency of the Peano arithmetics is represented by the adiabacity of the modeling thermodynamic system.

Peano Axioms/Inference Rules in the System $\mathcal{P}$/Theory $\mathcal{T}_{\mathcal{PA}}$.

$1/\mathcal{P}$ $\mathbb{N}_0 = \mathbb{N} \cup \{0\}$;

2 $\forall_{x \in \mathbb{N}_0}[[\exists_{y \in \mathbb{N}}|[y = f(x)]]$;

$3/\mathcal{P}$ $\forall_{x \in \mathbb{N}_0}|[0 \neq f(x)]$;

4 $\forall_{x \in \mathbb{N}_0}|[[f(x) \neq f(y)] \Rightarrow (x \neq y)]$;

$5/\mathcal{P}$ *axiom/axiomatic schema of the **mathematical induction**:*

$$[[\varphi(0) \wedge \forall_{x \in \mathbb{N}_0}|\varphi(x) \Rightarrow \varphi[f(x)]] \Rightarrow \forall_{x \in \mathbb{N}_0}|\varphi(x)]$$

Inference rule *Modus Ponens*

$$\frac{\vdash b, \ \vdash (b \Rightarrow c)}{\vdash c}, \quad c \text{ - } immediate \ consequence \ of \ b$$

Inference rule *Generalization*

$$\frac{a}{\forall_v a}, \ \frac{a}{x\Pi(a)}, \ \text{ better } \ \frac{\vdash a}{\vdash \forall_v a}, \ \frac{\vdash a}{\vdash [x\Pi(a)]}, \quad \forall_v a / x\Pi(a) \text{ - } immediate \ consequence \ of \ \mathbf{a}$$

♣ "1" - arithmeticity of the $\mathcal{P} \cong$ adiabacity of the $\mathcal{L}/\mathcal{D}_{\mathcal{L}}$.

♣ Consistent $\mathcal{T}_{\mathcal{PA}}$ inference within $\mathcal{P} \cong$ moving along trajectories $1_{\mathcal{D}\mathcal{L}}$ in $\mathcal{D}_{\mathcal{L}}/\mathcal{L}$.

♣ The states on the adiabatic trajectories, also irreversible, then model the consistently inferred/inferrable *PA-FORMULAS*.

Remark: Any *inference* within the system $\mathcal{P}^2$ sets the $\mathcal{T}_{\mathcal{PA}}$-*theoretical* relation[3] among its formulae $a_{[\cdot]}$. This relation is given by their gradually generated *special sequence* $\vec{a}\,[a_1, ..., a_q, ..., a_p, ..., a_k, a_{k+1}]$, which is the *proof* of the latest inferred formula a_{k+1}. By this, the *unique* arithmetic relation between their *Gödel numbers*, *FORMULAE* $x_{[\cdot]}$, $x_{[\cdot]} = \Phi(a_{[\cdot]})$, is set up, too. The gradually arising *SEQUENCE of FORMULAE* $x = \Phi\left(\vec{a}\right)$ is the *PROOF* of its latest *FORMULA* x_{k+1}.

Let us assume that the given sequence $\vec{a} = [a_{o1}, a_{o2}, ..., a_o, ..., a_q, ..., a_p, ..., a_k, a_{k+1}]$ is a special one, and that, except of axioms (axiomatic schemes) $a_{o1}, ..., a_o$, it has been generated by the correct application of the rule *Modus Ponens* only.[4]

Within the process of the *(Gödelian) arithmetic-syntactic analysis* of the latest formula a_{k+1} of the proof $\vec{a}$, we use, from the $\vec{a}$ *selected*, (special) subsequence $\overrightarrow{a_{q,p,k+1}}$ of the formulae a_q, a_p, a_{k+1}. The formulae a_q, a_p have already been derived, or they are axioms. It is valid that $q, p < k+1$, and we assume that $q < p$,

$$\overrightarrow{a_{q,p,k+1}} = [a_q, a_p, a_{k+1}], \quad a_p \cong a_q \supset a_{k+1}, \overrightarrow{a_{q,p,k+1}} = [a_q, \ a_q \supset a_{k+1}, a_{k+1}],$$

$$x = \Phi\left(\vec{a}\right) = \Phi\big([\Phi(a_1), \Phi(a_2), ..., \Phi(a_q), ..., \Phi(a_p), \ ..., \Phi(a_k), \Phi(a_{k+1})]\big)$$

$$= \Phi\left(\vec{x}\right) = \Phi(x_1) * \Phi(x_2) * ... * \Phi(x_q) * ... * \Phi(x_q) * ... * \Phi(x_k) * \Phi(x_{k+1})$$

$$l(x) = l\left[\Phi\left(\vec{x}\right)\right] = l\left[\Phi\left(\vec{a}\right)\right] = k + 1,$$

$$x_{k+1} = \Phi(a_{k+1}) = l\left[\Phi\left(\vec{a}\right)\right] Gl \, \Phi\left(\vec{a}\right) = (k+1) Gl \, x$$

$$x_p = \Phi(a_p) = \Phi(a_p \supset a_{k+1}) = q Gl \, \Phi\left(\vec{a}\right) * \Phi(\supset) * l\left[\Phi\left(\vec{a}\right)\right] Gl \, \Phi\left(\vec{a}\right)$$

$$= q Gl \, x Imp \, [l(x)] Gl \, x$$

$$x_q = \Phi(a_q) = q Gl \, \Phi\left(\vec{a}\right) = q Gl \, x$$

---

[2] Formal arithmetic inferential system.

[3] Peano Arithmetics Theory.

[4] For simplicity. The 'real' inference is applied to the formula a_{i+1} for $i = o$.

Checking the *syntactic and T_{PA}-theoretical correctness* of the analyzed chains a_i, as the formulae of the system P having been generated by inferring (*Modus Ponens*) within the system P (in the theory T_{PA}), and also the special sequence of the formulae $\vec{a}$ of the system (theory T_{PA}), is realized by checking the *arithmetic-syntactic* correctness of the notation of their corresponding *FORMULAE* and *SEQUENCE of FORMULAE*, by means of the relations $Form(\cdot)$, $FR(\cdot)$, $Op(\cdot,\cdot,\cdot)$, $Fl(\cdot,\cdot,\cdot)$ "called" from (the sequence of procedures) relations $Bew(\cdot)$, $(\cdot\cdot)B(\cdot)$, $Bw(\cdot)$;[5] the core of the whole (Gödelian) arithmetic-syntactic analysis is the (procedure) relation of *Divisibility*,

$$Form[\Phi(a_i)] = \; "1"/"0", \quad FR\left[\Phi\left(\overrightarrow{a_1^{i+1}}\right)\right] = \;"\; 1"/"0", \; o \leq i \leq k$$

$$Op[x_k, \; Neg(x_q), \; x_{k+1}] = Op[\Phi(a_p), \; \Phi[\sim(a_q)], \; \Phi(a_{k+1})] = \;"1"/"0"$$

$$Fl[(k+1)Glx, \; pGlx, \; qGlx] = \;"\; 1"/"0"$$

$$xBx_{k+1} = \;"1"/"0", \quad Bew(x_{k+1}) = \;"1"/"0";$$

$$\Phi(a_p)\|23^{3Gl\Phi\left(\overrightarrow{a_{q,p,k+1}}\right)} \quad \& \quad \mathbf{\Phi}(a_p)\|7^{1Gl\Phi\left(\overrightarrow{a_{q,p,k+1}}\right)} = \;"\; 1"/"0"$$

2. Gödel theorems

Remark: The expression $Sb\left(t \begin{array}{cc} u_1 & u_2 \\ Z(x) & Z(y) \end{array}\right)$ or the expression

$Sb\left(t \begin{array}{cc} 17 & 19 \\ Z(x) & Z(y) \end{array}\right)$ represents the result value of the Gödel number $t[Z(x), Z(y)]$, which is coding the (constant) claim $T(x, y)$ z *PM* has been generated by the substitution of x a y instead of the free variables X and Y in the function $T(X, Y)$ from *PM* with its Gödelian code $t(u_1, u_2)$ in the (arithmetized) P,

$$Sb\left(t \begin{array}{cc} u_1 & u_2 \\ Z(x) & Z(y) \end{array}\right) = Sb\left(t \begin{array}{cc} 17 & 19 \\ Z(x) & Z(y) \end{array}\right)$$

♣ **Into the *VARIABLES*, we substitute the *SIGNS* of the same *type* but the introduction of the term *admissible substitution* itself is not supposing it wordly.**
- Then it is possible to work even with the expressions not grammatically correct and thus with such chains, which are not *FORMULAE* of the system P (and thus not belonging into the theory T_{PA}).

Then the substitution function $Sb\left(\cdot \begin{array}{c} \cdots \\ \cdots \end{array}\right)$ is not possible, within the frame

of the inference in the system P, be used isolately as an arbitrarily performed number manipulation—in spite of the fact that it is such number manipulation really. It is used only and just within the frame of the language $\mathcal{L}_P$ and, above all, within the frame of the conditions specified by the právě a jenom

---

[5] *Formula, Reihe von Formeln, Operation, Folge, Glied, Beweis, Beweis*, see Definition 1–46 in [5–7] and by means of all other, by them 'called', relations and functions (by their procedures).

INFERENCE of the elements of the language $\mathcal{L}_{T_{PA}}$ only (and thus in the more limited way).

Others than/semantically (or by the type) homogenous application of the substitution function is not within the right inference/INFERENCE within the system $\mathcal{P}$ possible.[6]

2.1 The Gödel *UNDECIDABLE CLAIM*'s construction

◆ Let the Gödel numbers x and y be given. The number x is the *SEQUENCE OF FORMULAE* valid and y is a *FORMULA* of $\mathcal{P}$. We define the valid constant relation $Q(x, y)$ from the $Q(X, Y)$ for given values x and y, $X := x$, $Y := y$; $17 = \Phi(X)$, $19 = \Phi(Y)$,[7,8]

$$Q(x, y) \equiv xB_\kappa \left[Sb \left(\overline{\begin{matrix} 19 \\ y \\ Z(p) \end{matrix}} \right) \right] \equiv Bew_\kappa \left[Sb \left(\begin{matrix} 17 & 19 \\ q & \\ Z(x) & Z(y) \end{matrix} \right) \right]$$

$$q[Z(x), \ Z(y)] = \Phi[Q(x, \ y)], \ \overline{xB_\kappa y'} \equiv Bew_\kappa(y') = Bew_\kappa[y[Z(y)]] = Bew_\kappa[q[Z(x), \ Z(y)]]$$

(1)

◆ Now we put $p = 17Gen\, q$, $q = q(17, 19)\ [q(17,19) \triangleq Q(X, Y)]$ and then,

$$p \ = \ 17Gen\, q(17, 19) = \Phi[\forall_{x \in \mathbb{X}} | Q(x, Y)] \triangleq Q(\mathbb{X}, Y) \triangleq Q(\mathbb{N}_0, Y)$$

(2)

The meta-language symbol $Q(\mathbb{X}, Y)$ or $Q(\mathbb{N}_0, Y)$ is to be read: **No** $x \in \mathbb{X}(\mathbb{N}_0)$ **is in the κ-INFERENCE relation to the variable** Y (to its space of values $\mathbb{Y}$).

◆ Further, with the Gödel substitution function, we put $q[17, Z(p)] = r(17) = r$,

$$r := Sb \left(\begin{matrix} 19 \\ q \\ Z(p) \end{matrix} \right) \text{ and then } r = Sb \left(\begin{matrix} 19 \\ q(17, 19) \\ Z(p) \end{matrix} \right) = r(17) = \Phi[Q(X, \ p)] \quad (3)$$

The Gödel number r is, by the substitution of the *NUMERAL* $Z(p)$, **supposedly only** (by [5–7]) the *CLASS SIGN* with the *FREE VARIABLE* 17 (X); with the values p, the r contains the feature of *autoreference*,

$$r = r(17) = q[17, \ Z[p(19)]] = q[17, \ Z[17Gen\, q(17, 19)]] \triangleq Q(X, \ p)$$
$$= \Phi[Q[X, \ \Phi[\forall_{x \in X} | Q(x, \ Y)]]]_{Y := p} \triangleq Q[X, \ \Phi[Q(\mathbb{X}, \ Y)]] \triangleq Q[X, \ \Phi[Q(\mathbb{N}_0, \ Y)]]$$

(4)

◆ Within the Gödel number/code q, $q = q$ [17, 19], we perform the substitution $Y := p$ and then $X := x$ and write

---

[6] *Substitution function* $Sb \left(\begin{matrix} \cdots \\ \cdot \\ \cdots \end{matrix} \right)$ is, in this way, similar to the computer *machine instruction* which itself,

is always able to realize its operation with its operands on the arbitrary storage place, but practically it is always applicated within the limited *address space* and within the given *operation regime/mode* of the computer's activity only (e.g. regime/mode *Supervisor* or *User*).

[7] Φ and Z represents the *Gödel numbering* and Sb the *Substitution*, B, Bew the PA-arithmetic *Proof*.

[8] Following the Gödel Proposition V (the first part) [5–7].

$$r[Z(x)] = Sb \begin{pmatrix} 17 & 19 \\ q(17,19) & \\ & \\ Z(x) & Z(p) \end{pmatrix} = Sb \begin{pmatrix} & 17 & \\ q[17, Z(p)] & \\ & \\ & Z(x) \end{pmatrix}$$
$$= \Phi[Q[x, \ \Phi[Q(\mathbb{X}, \ Y)]]] = \Phi[Q[x, \ \Phi[Q(\mathbb{N}_0, \ Y)]]] = \Phi[Q(x, \ p)] \tag{5}$$

With the great quantification of $r[Z(x)]$ by $Z(x)$ by the *VARIABLE X* (17), we have (similarly as in [4, 8]),

$$Z(x) Gen\, r[Z(x)] = 17 Gen\, q[17, \ Z[17 Gen q(17,19)]] = 17 Gen\, r(17) = 17 Gen\, r$$
$$\triangleq \Phi[\forall_{x \in X} | \Phi[Q[x, \Phi[\forall_{x \in X} | Q(x, Y)]]]] = \Delta Q[\mathbb{X}, \Phi[Q(\mathbb{X}, Y)]] = Q[\mathbb{N}_0, \Phi[Q(\mathbb{N}_0, Y)]]$$
$$\tag{6}$$

2.2 Gödel theorems

I. **Gödel theorem** (corrected semantically by [3, 9, 10]) claims that
♣ **for every** *recursive* **and** *consistent CLASS OF FORMULAE* κ **and outside this set there is such true ("1")** *CLAIM r* **with free** *VARIABLE* $v\left[r \triangleq r(v)\right]$ **that neither** *PROPOSITION* $vGen\, r$ **nor** *PROPOSITION* $Neg(vGen\, r)$ **belongs to the set** $Flg(\kappa)$,

$$[vGen\, r \notin Flg(\kappa)] \quad \& \quad [Neg(vGen \ r) \notin Flg(\kappa)] \tag{7}$$

FORMULA $vGen\, r$ **and** $Neg(vGen\, r)$ **are not** κ-*PROVABLE*—*FORMULA* $vGen$ r *is not* κ-*DECIDABLE.* They both are elements of inconsistent (meta)system $\mathcal{P}^*$.
II. Gödel **theorem** (corrected semantically according to [3, 9, 10]) claims that
♣ **if** κ **is an arbitrary** *recursive* **and** *consistent CLASS OF FORMULAE,* **then any** *CLAIM* **saying that** *CLASS* κ **is consistent must be constructed outside this set, and for this fact it is not** κ-*PROVABLE.*
- Outside[9] the consistent system $\mathcal{P}_\kappa$, there is a true ("1") formula,[10] the *ARITHMETIZATION* of which is κ-*UNPROVABLE FORMULA* $17 Gen\, r$.[11]
♦ The fact that the recursive *CLASS OF FORMULAE* κ (now *PA—Peano Arithmetic* especially) is consistent, is tested by *unary* **relation** $Wid(\kappa)$, (die *Widerspruchsfreiheit, Consistency*) [5–7],

$$Wid(\kappa) \sim (Ex)\left[Form(x) \ \& \overline{Bew_\kappa(x)}\right] \tag{8}$$

- **a class of** *FORMULAE* κ **is consistent** $\overset{\text{Def}}{\Leftrightarrow}$ **there exists** *at least one FORMULA* x [*PROPOSITION* x ($x = 17 Gen\, r$)], **which is** κ-*UNPROVABLE.*

3. Caratheodory theorems

I. **Caratheodory's theorem** ($\Rightarrow$) says that: ◊ If the *Pfaff form has an integration factor,* **then there are,** *in the arbitrary vicinity of any arbitrarily chosen and fixed* point

---

[9] Far from (!) "In ...." in [5–7]

[10] Far from "... [*PA*-]arithmetic and sentencial/*SENTENCIAL*" in [5–7].

[11] Any attempt to prove/*TO PROVE* it (to infer/to *TO INFER* it) in the system $\mathcal{P}_\kappa$ assumes or leads to the requirement for inconsistency of the consistent (!) system $\mathcal{P}_\kappa$ (in fact we are entering into the inconsistent metasystem $\mathcal{P}^*$ - see the real sense [4, 9] of the Proposition *V* in [5–7]).

P of the *hyperplane* $\mathcal{R}$ $\left[P \in \mathcal{R}\left[(x_i)_{i=1}^n\right] = \text{const.}\right]$, **such points which**, from this point P, **are inaccessible** along the path satisfying the equation $dQ = 0$.

II. **Caratheodory theorem** ($\Leftarrow$) says that: $\lozenge$ *If the **Pfaff form** $\delta Q = \sum_{i=1}^n X_i dx_i$, where X_i are continuously differentiable functions of n variables* (over a simply continuous area), *has such a property that in the arbitrary vicinity of any arbitrarily chosen and fixed point P of the hyperplane* $\mathcal{R}$ $\left[P \in \mathcal{R}\left[(x_i)_{i=1}^n\right] = \text{const.}\right]$, **there exists such points which, from P, cannot be accessible along the path satisfying the equation** $dQ = 0$, *then this form is **holonomous**; it has or it is possible to find an integration factor for it.*

Caratheodory formulation of the *II*. Law of Thermodynamics ($\Leftrightarrow$) claims that:

$\lozenge$ *In the arbitrary vicinity of every state of the state space of the adiabatic system, **there are such states that, from the given starting point, cannot be reached along an adiabatic path*** (reversibly and irreversibly), *or such states which the system cannot reach at all*, see the **Figure 1**.

Remark: Now the symbol Q denotes that heat given to the state space of the thermodynamic system from its outside and directly; $Q \triangleq Q_{Ext}$; along paths l_{2b}, $l_{2b'}$, l_{2d}, l_{2e}, l_3 is $Q_{Ext} = 0$, $\Delta Q_{Ext} = 0$, $dQ_{Ext} = 0$.

♣ The states' $\theta_{[\cdot]}^{\mathcal{L}}$ changes in the adiabatic system $\mathcal{L}/\mathfrak{O}_{\mathcal{L}}$, along the trajectories $l_{\mathfrak{O}_{\mathcal{L}}}$ are expressible **regularly**:

$$
\begin{aligned}
&l_{2b} \quad \textbf{\textit{isothermic}} \;\; \textit{irreversible}, && \theta_1^{\mathcal{L}} \to \Delta A_{1,2e}\theta_{2e}^{\mathcal{L}} \\
&l_{2b'} \;\; \textbf{\textit{adiabatic}} \;\; \textit{irreversible}, && \theta_1^{\mathcal{L}} \to \Delta A_{1,3}\theta_3^{\mathcal{L}} \\
&l_{2d} \;\; \textbf{\textit{izobaric}} \;\; \textit{irreversible}, && \theta_1^{\mathcal{L}} \to \Delta A_{1,2b}\theta_{2b}^{\mathcal{L}} \\
&l_{2e} \;\; \textbf{\textit{izentropic}} \;\; \textit{reversible}, && \theta_1^{\mathcal{L}} \to \Delta A_{1,2b'}\theta_{2b'}^{\mathcal{L}} \\
&l_3 \;\; \textbf{\textit{izochoric}} \;\; \textit{irreversible}, && \theta_1^{\mathcal{L}} \to \Delta A_{1,2d}\theta_{2d}^{\mathcal{L}} \\
& && \theta_1^{\mathcal{L}} \to \lambda\theta_1^{\mathcal{L}}
\end{aligned}
$$

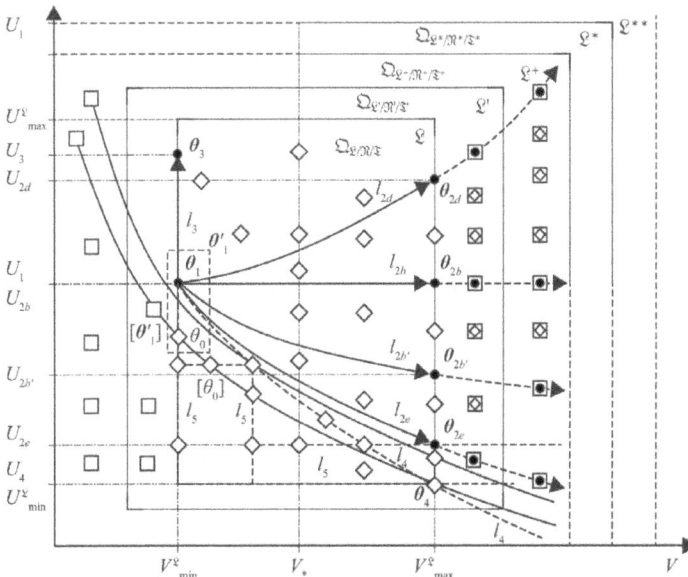

Figure 1.
Adiabatic changes of the state of the system $\mathcal{L}$, illustration.

Through the state space of *FORMULAE* of the system $\mathcal{P}$**, we "travel" simi-larly** by the inference rules, *Modus Ponens* especially [performed by a **Turing Machine *TM*,** the inference of which is considerable as realized by the *information transfer process* within a **Shannon Transfer Chain** $(\mathbf{X}, \mathcal{K}, \mathbf{Y})$ described thermody-namically by a **Carnot Machine *CM*].**

The thermodynamic model for the consistent $\mathcal{P}/\mathcal{T}_{\mathcal{PA}}$ **inference,** from its axioms or formulas having been inferred so far, **is created by the Carnot Machine's activity, which models the inference. This whole Carnot Machine *CM* runs in the wider adiabatic system** $\mathcal{L}/\mathfrak{D}_{\mathcal{L}}$ **and, in fact, is, in this way, creating these states,** [the *TM*'s, $(\mathbf{X}, \mathcal{K}, \mathbf{Y})$'s, configurations are then modeled by the states $\theta_i^{\mathcal{L}} \in \mathfrak{D}_{\mathcal{L}}$ of the adiabatic $\mathcal{L}/\mathfrak{D}_{\mathcal{L}}$ with this modeling *CM* inside], see the **Figure 2.**

The $\mathcal{L}$**'s initial imbalance starts the** $\theta_{[\cdot]}^{\mathcal{L}}$**s states' sequence on a trajectory** $l_{\mathfrak{D}_{\mathcal{L}}}$ and **is given by** the modeled

> temperature difference $T_W - T_0 > 0$ on CM,
>
> existence of the input message on $\mathcal{K}$,
>
> input chain's existence on the TM's input-output tape

These adiabatic trajectories $l_{\mathfrak{D}_{\mathcal{L}}}$ now **represent the norm** of the **consistency** (and resultativity) of the $\mathcal{P}/\mathcal{T}_{\mathcal{PA}}$**-inference/computing process** expressible also in terms of the **information transfer/heat energy transformation.**

♣ **The adiabatic property of the thermodynamic system** $\mathcal{L}$ **is always created** over the given scales of its state quantities—over their scale for a certain "creating" original (and not adiabatic) system $\mathfrak{T}$, and **by its *outerly specification* or *the design/ construction* by means of heat/adiabatic isolation of the space** V_{max} of the origi-**nal system** $\mathfrak{T}$ **that the system** $(\mathcal{L}/\mathfrak{T})$ **can occupy, and after the system** $\mathcal{L}$ **has been** (as the adiabatic isolated original system $\mathfrak{T}$) **designed and set in the starting state** θ_1, see **Figure 1. The state** θ_4 **is a state** $\lozenge$ **of the set of states** $\{\lozenge\}$**. These states are those ones in the Figure 1,** which, although they are in the given scale of state quantities U and V of the state space $\mathfrak{D}_{\mathcal{L}}$ of the system $\mathcal{L}$ considered, $U \in \langle U_{min}, U_{max} \rangle$ and $V \in \langle V_{min}, V_{max} \rangle$, are within it [in (the state space $\mathfrak{D}_{\mathcal{L}}$ of) $\mathcal{L}$]

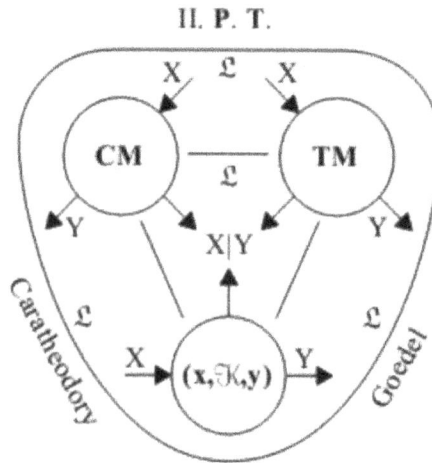

Figure 2.
*The mutual describability of the **CM**, $(\mathbf{X}, \mathcal{K}, \mathbf{Y})$ and **TM**.*

by permitted (adiabatic, $dQ_{Ext} = 0$) **changes** $l_{2b}, l_{2b'}, l_{2d}, l_{2e}$ and l_3, **inaccessible**. And certainly, thermodynamic states $\square$ beyond these scales, within the hierarchically higher systems, are not accessible from the inside of the system $\mathcal{L}/\mathfrak{T}$ itself, without its (not adiabatical) widening, either, see the **Figure 1.**

. Without violation of the adiabacity of the system $\mathcal{L}$, it is not possible to reach the state θ_4 from the state θ_1 along any simple path $l_{2b}, l_{2b'}, l_{2d}, l_{2e}$ in the state space $\mathfrak{D}_{\mathcal{L}}$,

♣ However, **outside the adiabacity of the system** $\mathcal{L}$ expressed by the relation $dQ_{Ext} = 0$, which means **under the opposite requirement** $dQ_{Ext} \neq 0$, **it is possible to design or to construct** a (nonadiabatic) **path linking a certain point/state of the** *state space* $\mathfrak{D}_{\mathcal{L}}$ **located, e.g., on** l_{2e} **with the point/state** θ_4; **for example, it is the path** l_4 **from** θ_1 **to** θ_4, now in a certain nonadiabatic system $\mathfrak{N}$, $\mathfrak{N} \subseteq \mathfrak{T}$ where, from the view of possibilities of changes of the state, see **Figure 1**, is valid that

$$\square \notin \mathfrak{D}_{\mathcal{L}/\mathfrak{T}}, \{\square\} \not\subseteq \mathfrak{D}_{\mathcal{L}/\mathfrak{T}}; \quad \Diamond \notin \{l_{2b}, l_{2b'}, l_{2d}, l_{2e}, l_3\}, \quad \{\Diamond\} \not\subseteq \mathfrak{D}_{\mathcal{L}/\mathfrak{T}}$$
$$\mathfrak{D}_{\mathcal{L}} = \mathfrak{D}_{\mathfrak{N}} = \mathfrak{D}_{\mathfrak{T}} \triangleq \mathfrak{D}_{\mathcal{L}/\mathfrak{T}} \quad \mathfrak{T} \supseteq \mathfrak{N} \not\supseteq \mathcal{L} \qquad (9)$$

. Further, it is possible to create for this nonadiabatic system $\mathfrak{N}$ an alternative adiabatic system $\mathcal{L}'(\mathfrak{D}_{\mathcal{L}'} \supseteq \mathfrak{D}_{\mathcal{L}})$ enabling adiabatic-isochoric changes, e.g., $\theta_{2e} \to \theta_4$.

.. Both the new adiabatic system $\mathcal{L}'$ and its nonadiabatic "model" $\mathfrak{N}$ can be a subsystem of another but also adiabatic and imminently superior system $\mathcal{L}^+$ having another/wider range of the state quantities than it was for the original systems $\mathcal{L}$ and $\mathfrak{N}$, $(\mathfrak{D}_{\mathcal{L}} \subseteq \mathfrak{D}_{\mathcal{L}'/\mathfrak{N}} \subseteq \mathfrak{D}_{\mathcal{L}^+})$. Then the path l_4 in the state space $\mathfrak{D}_{\mathcal{L}'/\mathfrak{N}}$ of the system $\mathcal{L}'/\mathfrak{N}$ will be, from the point of $\mathcal{L}'$ of the imminently superior adiabatic system $\mathcal{L}^+$, the adiabatic one—the system $\mathcal{L}'$ is already isolated in $\mathcal{L}^+$ and the system $\mathcal{L}^+$ itself is already created in a certain system $\mathcal{L}^*$ imminently superior to it, as an isolated/adiabatic substitute for the system $\mathfrak{N}'$ $\left(\mathfrak{D}_{\mathcal{L}'/\mathfrak{N}} \not\subseteq \mathfrak{D}_{\mathcal{L}^+}, \mathfrak{D}_{\mathcal{L}^+/\mathfrak{N}'} \not\subseteq \mathfrak{D}_{\mathcal{L}^*} \dots\right)$.

.. From the view of the *possibilities to change the state*, or from the view of the *energetic relations* (*8*), it is possible, see the **Figure 1**, to write,

$$\mathcal{L} \not\subseteq \mathcal{L}' \not\subseteq \mathcal{L}^+ \not\subseteq \mathcal{L}^* \not\subseteq \mathcal{L}^{**} \dots, \quad \mathfrak{N} \not\subseteq \mathfrak{N}' \not\subseteq \mathfrak{N}^+ \not\subseteq \mathfrak{N}^* \not\subseteq \mathfrak{N}^{**} \dots \qquad (10)$$

$\{\mathfrak{N}' \not\subseteq \mathcal{L}^+\}_{\mathcal{E}}$, $\mathfrak{N}$ is implemented in $\mathcal{L}'$, $\{\mathfrak{N}^+ \not\subseteq \mathcal{L}^*\}_{\mathcal{E}}$, $\mathfrak{N}'$ is implemented in $\mathcal{L}^+$ $\{\mathfrak{N}^+ \not\subseteq \mathcal{L}^*\}_{\mathcal{E}}$, $\mathfrak{N}^+$ is implemented in v $\mathcal{L}^*$, ...

We introduce a symbol $l_{\mathfrak{D}_{\mathcal{L}[\cdot]}}$ for adiabatic paths in the state spaces $\mathfrak{D}_{\mathcal{L}[\cdot]}$,

$$l_{\mathfrak{D}_{\mathcal{L}}} \triangleq \{l_{2b}, l_{2b'}, l_{2d}, l_{2e}, l_3\}, \quad l_{\mathfrak{D}_{\mathcal{L}'}} \triangleq \{l_{2b}, l_{2b'}, l_{2d}, l_{2e}, l_3, l_{2e} - l_{\theta_2, \theta_4}, l_5\}, \quad l_{\mathfrak{D}_{\mathcal{L}^+}}, l_{\mathfrak{D}_{\mathcal{L}^*}}, \dots$$
$$l_{\mathfrak{D}_{\mathcal{L}}} \not\subseteq l_{\mathfrak{D}_{\mathcal{L}'}} \not\subseteq l_{\mathfrak{D}_{\mathcal{L}^+}} \not\subseteq l_{\mathfrak{D}_{\mathcal{L}^*}} \not\subseteq \dots$$
$$(11)$$

♣ The states from the sets $\{\mathfrak{D}_{\mathcal{L}} - l_{\mathfrak{D}_{\mathcal{L}}}\}$, $\{\mathfrak{D}_{\mathcal{L}'} - l_{\mathfrak{D}_{\mathcal{L}}}\}$, $\{\mathfrak{D}_{\mathcal{L}^+} - l_{\mathfrak{D}_{\mathcal{L}}}\}$, $\{\mathfrak{D}_{\mathcal{L}^*} - l_{\mathfrak{D}_{\mathcal{L}}}\}$, ... in the view of adiabacity and specification of the system $\mathcal{L}$ are forming, within the hierarchy of the systems $\mathcal{L}$, $\mathcal{L}'$, $\mathcal{L}^+$, $\mathcal{L}^*$, ..., a certain set $\mathfrak{D}_{\mathcal{L}}^* = \{\{\Diamond\} \cup \{\square\}\}$, which is in the framework of the system $\mathcal{L}$ inaccessible/unachievable as a whole and also in any of its subset and member. However, the $\mathcal{L}$-*inaccessibility* (adiabatic inaccessibility, especially of $\{\Diamond\}$ in the state space $\mathfrak{D}_{\mathcal{L}/\mathfrak{T}}$) also means existence of the paths $l_{\mathfrak{D}_{\mathcal{L}}}$ of the adiabatic system $\mathcal{L}$. In the sense of the domain of solution of its (the $\mathcal{L}$'s) state equations, **they cannot be part of the functionality of** $\mathcal{L}$ (but mark it).

4. Analogy between adiabacity and *PA*-inference

♣ Now **the states on the adiabatic paths** $l_{\mathfrak{Q}_e}$ (of changes of the state of the adiabatic system $\mathfrak{L}$) are considered to be the analogues of *PA*-**arithmetic claims/ claims of the** *Peano Arithmetic* **theory** T_{PA} (formulated/inferred/proved in $\mathcal{P}$),
   - **adiabacity of the system** $\mathfrak{L}$ is the analog of **consistency** of the system $\mathcal{P}_{[\kappa]}$ and
   . **the set** $l_{\mathfrak{Q}_e}$ **of adiabatic paths in** $\mathfrak{Q}_{\mathfrak{L}/\mathfrak{T}}$ is an analog of *PA*-**theory** T_{PA}; then, adiabatic analogy of the higher consistent inferential system $\mathcal{P}'$ is by $\mathfrak{L}'$, $\mathcal{P}' \supsetneq \mathcal{P}$, ...
   .. Then the given specific **adiabatic path** l_{2b}, $l_{2b'}$, l_{2d}, l_{2e}, l_3 is an analog of certain *deducible thread* $\vec{x} Bx_k$ **of the claim** x_k **of the theory** T_{PA}, where

$$\vec{x} \; Bx_k = (x_1, \; x_2, \; \dots, \; x_{k-1}, \; x_k)Bx_k = \text{"1"}$$
$$x_1 \in \{AXIOMS\}^{\mathcal{P}} \quad \text{and} \quad x_1 \cong \theta_1$$
$$x_1, \; x_2, \; \dots, x_{k-1}, \; x_k \in T_{PA} \quad \text{and} \quad x_k \cong \theta \in \{\theta_{2b}, \; \theta_{2b'}, \; \theta_{2d}, \; \theta_{2e}, \; \theta_3\} \qquad (12)$$
$$x_2, \dots, x_{k-1} \quad \cong \quad \theta \in \{\{\{l_{2b} - \theta_{2b}\}, \; \{l_{2b'} - \theta_{2b'}\}, \; \{l_{2d} - \theta_{2d}\},$$
$$\{l_{2e} - \theta_{2e}\}, \; \{l_3 - \theta_3\}\} - \theta_1\}$$

♣ The **states from the space** $\mathfrak{Q}_{\mathfrak{L}/\mathfrak{T}}$ **of the system** $\mathfrak{L}/\mathfrak{T}$ satisfying the range of values of the state quantities $p \in \langle p_{min}, p_{max} \rangle$, $V \in \langle V_{min}, V_{max} \rangle$, $T \in \langle T_{min}, T_{max} \rangle /$ $U \in \langle U_{min}, U_{max} \rangle$), **which are inaccessible along any of the adiabatic paths from** $l_{\mathfrak{Q}_L}$, that means they are the states $\lozenge$ from the difference $\{\mathfrak{Q}_{\mathfrak{L}/\mathfrak{T}} - l_{\mathfrak{Q}_e}\}$, shortly said from $\{\mathfrak{T} - \mathfrak{L}\}$, are considered to be analogues of not *PA*-claims such as, e.g., the **Fermat's Last Theorem.**[12] So, they are **analogues of** *all-the-time* **true ("1") arithmetic but not-*PA*-arithmetic claims.** From the point of adiabacity of the system $\mathfrak{L}$, they ($\lozenge$) are only some thermodynamic states of its "creating" system $\mathfrak{T}$, and they are from the common range of values of the state quantities for $\mathfrak{T}$ and $\mathfrak{L}$. From the point of expressing possibilities it as always true

$$\mathfrak{L}^{[\cdot]} \subsetneq \mathfrak{N}^{[\cdot]} \subseteq \mathfrak{T}^{[\cdot]} \subsetneq \mathfrak{T}^\star \subsetneq \left\{\mathfrak{Q}_{\mathfrak{L}^{[\cdot]}/\mathfrak{T}^{[\cdot]}}\right\}^\star \qquad (13)$$

[Symbol $\mathfrak{T}^\star$ denotes thermodynamic theory as a whole and symbol $\left\{\mathfrak{Q}_{\mathfrak{L}^{[\cdot]}/\mathfrak{T}^{[\cdot]}}\right\}^\star$ is a mark for a transitive and reflexive closure of the set of (any) claims about systems $\mathfrak{L}^{[\cdot]}//\mathfrak{T}^{[\cdot]}$.].
   ♣ The whole set $\mathfrak{O}_{\mathfrak{L}}^\star$ of **states inaccessible in a given scale of state quantities** of the system $\mathfrak{L}/\mathfrak{T}$ along the arbitrary **adiabatic path from** $l_{\mathfrak{Q}_L}$ **in the system** $\mathfrak{L}$ (states $\lozenge$), as well as **the set of** $\mathfrak{L}$-**inaccessible states** $\square$ **outside this scale,** see **Figure 1,** are considered now to be the thermodynamic bearer of analogy of the **semantics of the Gödel's** *UNDECIDABLE PROPOSITION* 17Gen r,

$$17Gen\, r \cong \mathfrak{Q}_{\mathfrak{L}}^\star \quad [= \{\{\lozenge\} \cup \{\square\}\}]$$
$$\mathfrak{O}_{\mathfrak{L}}^\star = \{\mathfrak{Q}_{\mathfrak{L}^\bullet} - l_{\mathfrak{Q}_L}\}, \quad \mathfrak{Q}_{\mathfrak{L}}^\star \subsetneq \mathfrak{Q}_{\mathfrak{L}^\bullet} \quad \subsetneq \; \{\mathfrak{Q}_{\mathfrak{L}^\ast/\mathfrak{T}^\ast}\}^\star \qquad (14)$$

   - **The states from** $\mathfrak{O}_{\mathfrak{L}}^\star$ (from $\{\mathfrak{Q}_{\mathfrak{L}^\bullet} - l_{\mathfrak{Q}_L}\}$, $\{\mathfrak{Q}_{\mathfrak{L}^\bullet} - l_{\mathfrak{Q}_e}\}$, $\{\mathfrak{Q}_{\mathfrak{L}^\bullet} - l_{\mathfrak{Q}_{e'}+}\}$, ..., $\{\mathfrak{Q}_{\mathfrak{L}^{\bullet\bullet}} - l_{\mathfrak{Q}_e}\}$, ..., $\{\mathfrak{Q}_{\mathfrak{L}^{\bullet\bullet}} - l_{\mathfrak{Q}_{e^\ast}}\}$, ...) **inaccessible by permitted changes in currently used systems** $\mathfrak{L}$, $\mathfrak{L}'$, $\mathfrak{L}^+$, $\mathfrak{L}^\ast$, ... (within the scale of values of their state quantities and also out of this scale) **confirm both existence and properties of**

---

[12] Alternatively Goldbach's conjecture.

these systems $\mathfrak{L}$, $\mathfrak{L}'$, $\mathfrak{L}^+$, $\mathfrak{L}^*$, ...; they confirm adiabacity of changes $l_{\mathfrak{Q}_\mathfrak{L}}$, $l_{\mathfrak{Q}_{\mathfrak{L}'}}$, $l_{\mathfrak{Q}_{\mathfrak{L}+}}$, $l_{\mathfrak{Q}_{\mathfrak{L}^*}}$, ... **running in them.**

For (to illustrate our analogy) a supposedly countable set of states along the paths $l_{\mathfrak{Q}_\mathfrak{L}}$ of changes of the state of the system $\mathfrak{L}$ (for simplicity we can consider the *isentrop* l_{2e} only), the *PROPOSITION* 17*Gen r* is a claim of countability set nature, the analog $\mathfrak{O}_\mathfrak{L}^*$ of which is formulated in the set $\left\{ \mathfrak{Q}_{\mathfrak{L}/\mathfrak{I}} \right\}^\star$; it as valid that

$$\{\mathfrak{Q}_{\mathfrak{L}/\mathfrak{I}}\}^\star \supsetneqq \mathfrak{Q}_\mathfrak{L} \quad \text{and} \quad \{\mathfrak{Q}_\mathfrak{L} - l_{\mathfrak{Q}_\mathfrak{L}}\} \subsetneqq \mathfrak{Q}_\mathfrak{L}^* \subsetneqq \{\mathfrak{Q}_{\mathfrak{L}/\mathfrak{I}}\}^\star$$
$$\{\mathfrak{Q}_\mathfrak{L} - l_{\mathfrak{Q}_\mathfrak{L}}\} \not\supseteq \mathfrak{Q}_\mathfrak{L} \quad \text{and} \quad \{\mathfrak{Q}_\mathfrak{L} - l_{\mathfrak{Q}_\mathfrak{L}}\} \not\supseteq l_{\mathfrak{Q}_\mathfrak{L}} \tag{15}$$

4.1 Analogy between Caratheodory and Gödel theorems

We claim that, *II.* **Caratheodory theorem,**

◊ *if an arbitrary* **Pfaff form** $\delta Q_{Ext} = \sum_{i=1}^n X_i dx_i$, *where X_i are functions of n variables, continuously differentiable* (over a simply continuous domain) *has such a quality that in the arbitrary vicinity of arbitrarily chosen fixed point P of the hyperplane* $\mathcal{R}\left[P \in \mathcal{R}, \mathcal{R}\left[(x_i)_{i=1}^n \right] = C = \text{const.} \right]$ *there exists* a set of points *inaccessible from the point P along the path satisfying the equation* $dQ_{Ext} = 0$, *then it is possible to find an integration factor for it and then this form is holonomous.* In a physical sense and, by means of the Thermodynamics language,

$$(\exists | l_{\mathfrak{Q}_L}) \Rightarrow (\exists | \mathfrak{Q}_\mathfrak{L}) \Rightarrow (\exists | \mathfrak{Q}_\mathfrak{L}^*) \Rightarrow (\exists | \mathfrak{Q}_{\mathfrak{L}^*}) \Rightarrow \left(\exists | \{\mathfrak{Q}_{\mathfrak{L}/\mathfrak{I}}\}^\star \right) \tag{16}$$

it says what, in its consequence [*w* 17*Gen r*, (8)] and in a meta-arithmetic-logical way, the *II.* **Gödel theorem** (corrected semantically by [3, 9, 10]) **claims;**

♣ if κ is an arbitrary *recursive* and *consistent CLASS OF FORMULAE*, then any *CLAIM* (written as the *SENTENCIAL* and as such, representing a countable set of claims, which are its implementations) saying that *CLASS* κ is consistent must be constructed outside this set and for this fact it is not κ-*PROVABLE*/is κ-*UNPROVABLE* or cannot be κ-*PROVABLE*. In fact, it is a part of the inconsistent metasystem $\mathcal{P}^*$.

- Outside the consistent system $\mathcal{P}_\kappa$, there is a true ("**1**") formula whose *ARITHMETIZATION* is κ-*UNPROVABLE FORMULA/PROPOSITION/CLAIM* or code 17*Gen r*".[13]

. In a physical sense and by the Thermodynamics language,

$$\{\mathfrak{Q}_{\mathfrak{L}/\mathfrak{I}}\}^\star \not\subseteq \mathfrak{Q}_{\mathfrak{L}^*} \not\subseteq \mathfrak{Q}_\mathfrak{L}^* \not\subseteq \mathfrak{Q}_\mathfrak{L} \not\subseteq \mathfrak{Q}_\mathfrak{L} \tag{17}$$

♣ It is possible to claim that, *I.* **Caratheodory theorem,**

◊ *if an arbitrary* **Pfaff form** $\delta Q_{Ext} = \sum_{i=1}^n X_i dx_i$ *has a integration factor, then there are in the arbitrary vicinity of an arbitrarily chosen fixed point P of the hyperplane* $\mathcal{R}$ *some points inaccessible from this point* $P\left[P \in \mathcal{R}\left[(x_i)_{i=1}^n \right] = \text{const.} \right]$ *along the path satisfying the equation* $dQ_{Ext} = 0$. In a physical sense and by means of the Thermodynamics language,

$$[\mathfrak{Q}_\mathfrak{L}^* \not\subseteq \mathfrak{Q}_\mathfrak{L}] \wedge \left[\left(\{\mathfrak{Q}_{\mathfrak{L}/\mathfrak{I}}\}^\star - \mathfrak{Q}_\mathfrak{L}^* \right) \not\subseteq \mathfrak{Q}_\mathfrak{L} \right] \tag{18}$$

---

[13] Any attempt to prove/*TO PROVE* it (to infer/*TO INFER* it) within the system $\mathcal{P}_k$ assumes or leads to the *Circulus Viciosus*.

it says what, in a meta-arithmetic-logical way, the *I.* **Gödel theorem** (corrected semantically by [3, 9, 10]) **claims**;

♣ for every *recursive* and *consistent CLASS OF FORMULAE* κ and outside this set, there is such a true ("**1**") *CLAIM r* with free *VARIABLE v* $\left[r \triangleq r(v)\right]$ that neither *PROPOSITION vGen r* nor *PROPOSITION Neg(vGen r) belongs* to the set $Flg(\kappa)$,

$$[vGen\, r \notin Flg(\kappa)] \quad \& \quad [Neg(vGen\ r) \notin Flg(\kappa)] \tag{19}$$

FORMULA vGen r and *Neg(vGen r)* are not κ-*PROVABLE*—*FORMULA vGen r i s not* κ-*DECIDABLE*. They are elements of inconsistent (meta)system $\mathcal{P}^*$.

♣ **For us, as an isolated system** $\mathfrak{L}$, **to achieve such a "state,"** it is necessary to consider the states with values of state quantities which are not a part of the domain of solution of the state equation for $\mathfrak{L}$. The system $\mathfrak{L}$ has not been designed for them (so, we are facing inconsistency). For example, **the required volume** V **and temperature** T **should be greater than their maxima** V_{max} and T_{max} achievable by the system $\mathfrak{L}$. In order "to achieve" them, **the system** $\mathfrak{L}$ **itself would have to "get out of itself,"** and in order to obtain values V and T **greater** than V_{max} and T_{max}, **it would have to "redesign"/reconstruct itself.** However, it is **us, being in a position of the hierarchically higher object,** who has to do so, from the outside the state space $\mathfrak{Q}_{\mathfrak{L}/\mathfrak{T}}$ (from the outside the volume V_{max}), which the system may occupy now.[14]

- This "procedure" corresponds to the ***CLAIM/PROPOSITION/FORMULA*** 17*Gen r* construction by means of (Cantor's) **diagonal argument** and **Caratheodory proof**.

♣ The states unachievable within the state spaces of the systems $\mathfrak{L}$, $\mathfrak{L}'$, $\mathfrak{L}^+$, $\mathfrak{L}^*$, ... or inaccessible from them are creating, as a whole, a *certain class of equivalence or macrostate* $\mathfrak{O}^*_{\mathfrak{L}^{[\cdot]}}$ $\left(\mathfrak{Q}^*_{\mathfrak{L}^{[\cdot]}} \not\subseteq \mathfrak{Q}_{\mathfrak{L}^{[\cdot]}} \not\subseteq \mathcal{U}_{\mathfrak{Q}_{\mathfrak{L}^{[\cdot]}}}\right)$ in hierarchy of the state spaces, from the point of their possible development, of always superior systems $—\{\mathfrak{L}' \cup \mathfrak{L}^+ \cup \mathfrak{L}^*\}$ for $\mathfrak{L}$, $\{\mathfrak{L}^+ \cup \mathfrak{L}^*\}$ for $\mathfrak{L}'$, $\mathfrak{L}^*$ for $\mathfrak{L}^+$, ... The existence of the macrostate $\mathfrak{O}^*_{\mathfrak{L}^{[\cdot]}}$, already beginning from the original system $\mathfrak{L}$ (macrostate $\mathfrak{O}^*_{\mathfrak{L}}$), confirms the existence of the currently considered (adiabatic) system $\mathfrak{L}^{[\cdot]}$ and its properties, especially its adiabacity. And by this, in our analogy, it also con rms the consistency of its arithmetic/mathematical analog $\mathcal{P}, \mathcal{P}', \mathcal{P}^+$, ... (a complement of the set cannot exist without this set) and, on the contrary,

$$\left(\exists|\{\mathfrak{Q}_{\mathfrak{L}/\mathfrak{T}}\}^\star\right) \quad \Rightarrow \quad \left(\exists|\mathfrak{Q}^*_{\mathfrak{L}^{[\cdot]}}\right), \quad \left(\exists|\{\mathfrak{Q}_{\mathfrak{L}/\mathfrak{T}}\}^\star\right) \quad \Leftarrow \quad \left(\exists|\mathfrak{Q}^*_{\mathfrak{L}^{[\cdot]}}\right);$$
$$\left[\left[\left(\exists|\mathfrak{Q}^*_{\mathfrak{L}^{[\cdot]}}\right) \Rightarrow \left(\exists|\mathfrak{L}^{[\cdot]}\right)\right] \quad \& \quad \left[\left(\exists|\mathfrak{L}^{[\cdot]}\right) \Rightarrow \left(\exists|\mathfrak{Q}^*_{\mathfrak{L}^{[\cdot]}}\right)\right]\right] \tag{20}$$
$$\Rightarrow$$
$$\left[\left(\exists|\mathfrak{Q}^*_{\mathfrak{L}^{[\cdot]}}\right) \Leftrightarrow \left(\exists|\mathfrak{L}^{[\cdot]}\right)\right]$$

- Based just upon this point of view, we assign the set/macrostate or equivalence class $\mathfrak{O}^*_{\mathfrak{L}^{[\cdot]}}$ the meaning of the bearer of the sense of the **Gödel's** *UNDECIDABLE PROPOSITION* 17*Gen r* **for** $\mathcal{P}$,

---

[14] This also involves introduction of the representative θ_0 of Fermat's Last Theorem provided we are speaking about $\mathfrak{L}$ with $l_{\mathfrak{O}_\mathfrak{L}}$ and provided we require enlargement $\mathfrak{L}'$ in order to get $\mathfrak{L}' \cong \mathcal{P}'$. The specific states accessible in the state space $\mathfrak{O}_\mathfrak{L} = \{p \in \langle p_{min}, p_{max} \rangle, V \in \langle V_{min}, V_{max} \rangle,$ $T \in \langle T_{min}, T_{max} \rangle / U \in \langle U_{min}, U_{max} \rangle, ...$ of the isolated system $\mathfrak{L}$ through reversible or irreversible changes other than adiabatic are thermodynamic analogy (interpretation) of the enlargement of the axiomatics of the original system $\mathcal{P}_{[\kappa]}$ to the new system $\mathcal{P}', \mathcal{P}^+, ...,$ *similar/relative* to the $\mathcal{P}_{[\kappa]}$. Such an enlargement of the system $\mathcal{P}$ to a certain system $\mathcal{P}^{[\cdot]}$ enabled Andrew Wiles to prove the Fermat's Last Theorem. Through its representative θ_0 we enlarge $\mathfrak{L}$ to $\mathfrak{L}'$, $\mathfrak{L}' \cong \mathcal{P}'$.

- the $\mathcal{L}$-*unachievability* of the set $\mathfrak{O}^*_{\mathfrak{L}[\cdot]}$ is in the position of the analog for this, in fact, methodological axiom which has been formulated in a certain hierarchically higher inferential (meta)system $\mathcal{P}^*$, $\mathcal{P}^* \cong \{\mathfrak{O}_\mathfrak{L}\}^*$. In accordance with the above and with **Figure 1**, we write for $\mathcal{L}/\mathcal{P}$

$$\theta_4 \in \{\mathfrak{Q}_{\mathfrak{L}^*} - l_{\mathfrak{Q}_\mathfrak{L}}\} = \{\{\Diamond\} \cup \{\Box\}\} = \mathfrak{Q}^*_\mathfrak{L} \subsetneqq \{\{\mathfrak{Q}_{\mathfrak{L}^*}\}^* - l_{\mathfrak{Q}_\mathfrak{L}}\} \subsetneqq \{\mathfrak{Q}_\mathfrak{L}\}^* \qquad (21)$$

$$[l_{\mathfrak{Q}_\mathfrak{L}} \nvdash \theta_4] \Rightarrow [l_{\mathfrak{Q}_\mathfrak{L}} \nvdash \{\mathfrak{Q}_{\mathfrak{L}^*} - l_{\mathfrak{Q}_\mathfrak{L}}\}] \quad [\in \{\{\mathfrak{Q}_{\mathfrak{L}^*}\}^* - l_{\mathfrak{Q}_\mathfrak{L}}\}]$$

$$[l_{\mathfrak{Q}_\mathfrak{L}} \nvdash \{\mathfrak{Q}_{\mathfrak{L}^*} - l_{\mathfrak{Q}_\mathfrak{L}}\}] \cong [l_{\mathfrak{Q}_\mathfrak{L}} \nsubseteq \{\mathfrak{Q}_{\mathfrak{L}^*} - l_{\mathfrak{Q}_\mathfrak{L}}\}]$$

$$[l_{\mathfrak{Q}_\mathfrak{L}} \nvdash [l_{\mathfrak{Q}_\mathfrak{L}} \nvdash \{\mathfrak{Q}_{\mathfrak{L}^*} - l_{\mathfrak{Q}_\mathfrak{L}}\}]] \quad [\in \{\{\mathfrak{Q}_{\mathfrak{L}^*}\}^* - l_{\mathfrak{Q}_\mathfrak{L}}\}]$$

$$[l_{\mathfrak{Q}_\mathfrak{L}} \nvdash [l_{\mathfrak{Q}_\mathfrak{L}} \nvdash [l_{\mathfrak{Q}_\mathfrak{L}} \nvdash \{\mathfrak{Q}_{\mathfrak{L}^*} - l_{\mathfrak{Q}_\mathfrak{L}}\}]]] \quad [\in \{\{\mathfrak{Q}_{\mathfrak{L}^*}\}^* - l_{\mathfrak{Q}_\mathfrak{L}}\}], \quad \dots$$

and further, for the theory $l_{\mathfrak{O}_\mathfrak{L}}/\mathcal{T}_{\mathcal{PA}}$, following (1)–(6) and [4], we write

$$l_{\mathfrak{Q}_\mathfrak{L}} \cong \mathcal{T}_{\mathcal{PA}}, \quad card \; l_{\mathfrak{Q}_\mathfrak{L}} = card \; \mathcal{T}_{\mathcal{PA}} = \aleph_0$$

$$card \{\mathfrak{Q}_{\mathfrak{L}^*}\}^* = \aleph_1, \quad card \{\{\mathfrak{Q}_{\mathfrak{L}^*}\}^* - l_{\mathfrak{Q}_\mathfrak{L}}\} = 1$$

$$l_{\mathfrak{Q}_\mathfrak{L}} \cong 17, \; \{\{\Diamond\} \cup \{\Box\}\} \cong 19 \; [19 \in \{\{\mathfrak{Q}_{\mathfrak{L}^*}\}^* - l_{\mathfrak{Q}_\mathfrak{L}}\}]$$

$$y = q[17, 19] \cong [\theta_{[\cdot]} \nvdash \{\{\Diamond\} \cup \{\Box\}\}], \quad p = 17 Gen \, q[17, 19] \qquad (22)$$

$$y[Z(y)] = q[17, \; y] \cong [\theta_{[\cdot]} \nvdash [\theta_{[\cdot]} \nvdash \{\{\Diamond\} \cup \{\Box\}\}]]$$

$$[[\theta_{[\cdot]} \nvdash (\theta \cup \Box)] \in \{\{\theta\} \cup \{\Box\}\}^*]$$

$$[\theta_{[\cdot]} \nvdash [\theta_{[\cdot]} \nvdash \{\{\Diamond\} \cup \{\Box\}\}]] \quad [\in \{\{\Diamond\} \cup \{\Box\}\}^*]$$

$$[\theta_{[\cdot]} \nvdash [\theta_{[\cdot]} \nvdash [\theta_{[\cdot]} \nvdash \{\{\Diamond\} \cup \{\Box\}\}]]], \quad \dots$$

For 19: $= Z(p)$ is $p[Z(p)] = r(17)$ and $r(17) \cong \{\{\Diamond\} \cup \{\Box\}\}^*$ and so we can write neatly

$$\left[\forall_{\theta_{[\cdot]} \in l_{\mathfrak{Q}_\mathfrak{L}}}\right] \cong 17 Gen, \quad \left[\forall_{\theta_{[\cdot]} \in l_{\mathfrak{Q}_\mathfrak{L}}} | [\theta_{[\cdot]} \nvdash \{\{\Diamond\} \cup \{\Box\}\}]\right] \cong [17 Gen \, [q(17, 19)]]$$

$$[17 Gen \, r] \cong \left[\forall_{\theta_{[\cdot]} \in l_{\mathfrak{Q}_\mathfrak{L}}} | \theta_{[\cdot]} \nvdash \left[\forall_{\theta_{[\cdot]} \in l_{\mathfrak{Q}_\mathfrak{L}}} | \theta_{[\cdot]} \nvdash \{\{\Diamond\} \cup \{\Box\}\}\right]\right]$$

$$[17 Gen \, r] \cong [l_{\mathfrak{Q}_\mathfrak{L}} \nvdash [l_{\mathfrak{Q}_\mathfrak{L}} \nvdash \{\mathfrak{O}_{\mathfrak{L}^*} - l_{\mathfrak{Q}_\mathfrak{L}}\}]], \quad \dots$$

$$(23)$$

which is the same as (21).

- It is obvious from our thermodynamic analogy that *CLAIM/PROPOSITION* 17*Gen r* for has to be true and **in connection with Gödel's II. theorem, and in accordance with Caratheodory we claim** that

$$\mathfrak{O}^*_{\mathfrak{L}[\cdot]} \cong 17 Gen \, r \; \text{for} \; \mathcal{L}/\mathcal{P} \; [\nu Gen \, r \; \text{for} \; \mathcal{L}'/\mathcal{P}', \; \mathfrak{L}^+, \dots] \qquad (24)$$

♣ The notation 17*Gen r* itself expresses the property of the system $\mathcal{P}$ and also the theory $\mathcal{T}_{\mathcal{PA}}$, just as an *subject* which itself is not and cannot be the object of its own, and thus its notation **is not and cannot be one of the objects of the system** $\mathcal{P}$ [similarly, as (17) is valid, $\mathfrak{O}^*_\mathfrak{L} \nsubseteq \mathfrak{Q}_\mathfrak{L} \nsubseteq l_{\mathfrak{Q}_\mathfrak{L}}$].

Demonstration: Following (8) $[Wid(\mathcal{P}) \Rightarrow 17Gen\ r]$, we claim for the systems $\mathcal{L}/\mathcal{P}$ that

$$[dQ^{\mathcal{L}}_{Ext} = 0] \cong w, \quad [dQ^{\mathcal{L}}_{Ext} = 0] \equiv \mathcal{L}; \quad w \cong [dQ^{\mathcal{L}}_{Ext} = 0], \quad \mathcal{L} \equiv [dQ^{\mathcal{L}}_{Ext} = 0]$$

$$(\exists|\mathcal{L}) \Rightarrow (\exists|\mathcal{O}^*_{\mathcal{L}}), (\exists|\mathcal{O}^*_{\mathcal{L}}) \cong 17Gen\ r; \quad (\exists|\mathcal{O}^*_{\mathcal{L}}) \Rightarrow (\exists|\mathcal{L}), (\exists|\mathcal{L}) \cong 17Gen\ r$$

$$[(\exists|\mathcal{L}) \Rightarrow (\exists|\mathcal{O}^*_{\mathcal{L}})] \cong [w \Rightarrow (17Gen\ r)]; \quad [(\exists|\mathcal{O}^*_{\mathcal{L}}) \Rightarrow (\exists|\mathcal{L})] \cong [(17Gen\ r) \Rightarrow w]$$

so, that $\quad [[(\exists|\mathcal{O}^*_{\mathcal{L}}) \Rightarrow (\exists|\mathcal{L})] \cong [(17Gen\ r) \Rightarrow w]]$

&

$[[(\exists|\mathcal{L}) \Rightarrow (\exists|\mathcal{O}^*_{\mathcal{L}})] \cong [w \Rightarrow (17Gen\ r)]]$ and then

$(\exists|\mathcal{O}^*_{\mathcal{L}}) \equiv 17\ Gen\ r$

$$(25)$$

I. Gödel theorem (corrected semantically by [3, 9, 10]):

For every *recursive* and consistent *CLASS OF FORMULAE* κ, and outside this set, there exists the true ("1") *CLAIM* r with a free *VARIABLE* v that neither the *CLAIM* vGen r nor the *CLAIM* Neg(vGen r) *belongs* to the set Flg(κ)

$$[vGen\ r \notin Flg(\kappa)] \ \& \ [Neg(vGen\ r) \notin /Flg(\kappa)],$$

***CLAIMS* vGen r and Neg(vGen r) are not κ-*PROVABLE*, the *CLAIM* vGen r is not κ-*DECIDABLE*.**

[They are elements of the formulating/syntactic metasystem $\kappa^{\star}$, inconsistent against κ].

II. Gödel theorem (corrected semantically by [3, 9, 10]):

If κ is an arbitrary *recursive* and *consistent CLASS OF FORMULAE,* then any *CLAIM* saying that *CLASS* κ is consistent must be constructed outside this set and for this fact, it **is not κ-*PROVABLE*.**

The **consistency** of the *CLASS OF FORMULAE* κ is **tested** by the *relation* **Wid**(κ).

$$\mathbf{Wid}(\kappa) \ \sim \ (\mathbf{Ex})[\mathit{CLAIM}(\mathbf{x}) \ \& \ \overline{\mathrm{Proof}_{\kappa}(\mathbf{x})}]$$

The FORMULAE class κ is consistent.
$\Leftrightarrow$
at least one κ-UNPROVABLE CLAIM x exists.
Now x = 17Gen r $\notin \mathcal{P}/\mathcal{T}_{\mathcal{PA}}, \kappa = \mathcal{T}_{\mathcal{PA}}, \mathcal{T}_{\mathcal{PA}} \subset \mathcal{P} \subset \mathcal{P}^{\star}$

Then, **semantically understood** and with the language of logic and meta-arithmetics, the full meaning of the **Gödel proof expresses the universal validity of the *II.* Law of Thermodynamics.**[15]

$$[II.\mathrm{P.T.}[\mathrm{Proof}_{\mathcal{P}}(\underline{17Gen\ r}) = "0"] = "1"\ ] = \mathrm{Wid}(\mathcal{T}_{\mathcal{PA}})$$

---

[15] Our consideration is based on the similarity between the Cantor diagonal argument used in construction of the Gödel Undecidable Formula and the proof way of the Caratheodory theorems; adiabacity/consistency is prooved by leaving them and sustaining their validity - paradox.

5. Conclusion

Peano Arithmetic theory is generated by its inferential rules (rules of the inferential system in which it is formulated). It consists of parts bound mutually just by these rules, but none of them is not identical with it nor with the system in their totality.

By information-thermodynamic and computing analysis of Peano arithmetic proving, we have showed why the Gödel formula and its negation are not provable and decidable within it. They are constructed, not inferred, by the diagonal argument, which is not from the set of the inferential rules of the system. The attempt to prove them leads to awaiting of the end of the infinite cycle being generated by the application of the substitution function just by the diagonal argument. For this case, the substitution function is not countable and for this it is not recursive (although in the Gödel original definition is claimed that it is). We redefine it to be total by the zero value for this case. This new substitution function generates the Gödel numbers of chains, which are not only satisfying the recursive grammar of formulae but it itself is recursive. The option of the zero value follows also from the vision of the inferential process as it would be the information transfer. The attempt to prove the Gödel Undecidable Formula is the attempt of the transfer of that information, which is equal to the information expressing the inner structure of the information transfer channel. In the thermodynamic point of view, we achieve the equilibrium status, which is an equivalent to the inconsistent theory. So, we can see that the Gödel Undecidable Formula is not a formula of the Peano Arithmetics and, also, that it is not an arithmetical claim at all. From the thermodynamic consideration follows that even we need a certain effort or energy to construct it, within the frame of the theory this is irrelevant. It is the error in the inference and cannot be part of the theory and also it is not the system. Its information value in it (as in the system of the information transfer) is zero. But it is the true claim about **inferential properties of the theory** (in fact, of the **properties of the information transfer**).

Any description of real objects, no matter how precise, is only a model of them, of their properties and relations, making them available in a specified and somewhat limited (compared with the reality) point of view determined by the description/model designer. This determination is expressed in definitions and axiomatics of this description/model/theory—both with definitions and by axioms and their number. Hence, realistically/empirically or rationally, it will also be true about (objects of) reality what such a model, called *recursive and able-of-axiomatization*, does not include. With regard of reality any such a model is *axiomatically incomplete*, even if the system of axioms *is complete*. **In addition, and more importantly, this description/model of objects, of their properties and possible relations** (the theory about reality) **cannot include a description of itself** just as the object of reality defined by itself (any such theory/object is not a subject of a direct description of itself). The description/model or the theory about reality is a grammar construction with substitutes and axiomatization and, as such, it is *incomplete in the Gödelian way*—**the grammar itself does not prevent a semantical mixing; but any observed real object cannot be the subject of observation of itself and this is valid for the considered theory, just as for the object of reality, too**. No description of reality arranged from its inside or created within the theory of this reality can capture the reality completely in wholeness of its all own properties. It is impossible for the models/theories considered, independently on their axiomatization. They are limited in principle [in the *real sense* of the Gödel theorems (in the Gödelian way)].

Now, **with our better comprehension**, we can claim that the **consistency of the recursive and axiomatizable system can never be proved in it itself,** even if the system is consistent really. The reason is that a **claim of the consistency of such a system is designable only if the system is the object of** *outer observation/measuring/studies*, **which is not possible within the system itself.** Ignoring this approach is also the reason for the formulation of the **Gibbs paradox** and **Halting Problem.** Also, our awareness of this fact results in our **full understanding** of the **meaning and proof of the Gödel theorems**, very often explained and described incomprehensibly, even inconsistently or paradoxically, **and which is parallel with the way of the Caratheodory proof of the** *II.* **Thermodynamic Principle**.

> Vienna Circle, 1931-1935, logical positivists **Rudolph Carnap**
> and Otto von Neurath:
>
> *"Any scientifically meaningful statement is expressible in physical terms,* about a movement in the observable space and time *or,* if the statement is not expressible this way it is meaningful scientifically *when it is convertible to a statement about a language,* otherwise it is of no scientific meaning."

♣♦◊

Acknowledgements

Supported by the grant of Ministry of Education of the Czech Republic MSM 6046137307.

Many thanks are to be expressed to my brother Ing. Petr Hejna for his help with English language and formulations of both this and all the previous texts.

A. Appendix

A.1 Summarizing comparison

♣ **Under the adiabacity,** $[d]Q_{Ext} = 0$, **of the system** $\mathcal{L}$, *it is not possible to derive* **such a** *CLAIM* **that is stating this adiabatic supposition.** This *CLAIM* is constructible not adiabatically, *outside* the adiabatic $\mathcal{L}$ only.

♣ **Under the consistency of the system** $\mathcal{P}$, *it is not possible to derive* **such a** *CLAIM* **that is stating this consistency supposition.** This *CLAIM* is constructible *purely syntactically*, *outside* the consistent $\mathcal{P}$ only (in $\mathcal{P}^* - \mathcal{P}$) (Figure A1).

♣ **Without** $\mathcal{P}^*$ **we could not know that P is not self-referencing and is consistent.**

> Autoreference / *HALTING PROBLEM* / Self-Observation
>
> - the *CLAIM* about adiabacity of $\mathcal{L}$ within $\mathcal{L}$ -
>
> - the *CLAIM* about consistency of $\mathcal{T}_{PA}$ within $\mathcal{P}$ -
>
> is excluded.
>
> This is the nature law expressed by the **Caratheodory form** of the *II*. **P.T.** and by the **Gödel theorems' sense.**
>
> The eye can not look at and into itself.
>
> Any mixing of the various observation/expressing/approach levels leads to the paradoxes and is to be excluded from the cognitive thinking.

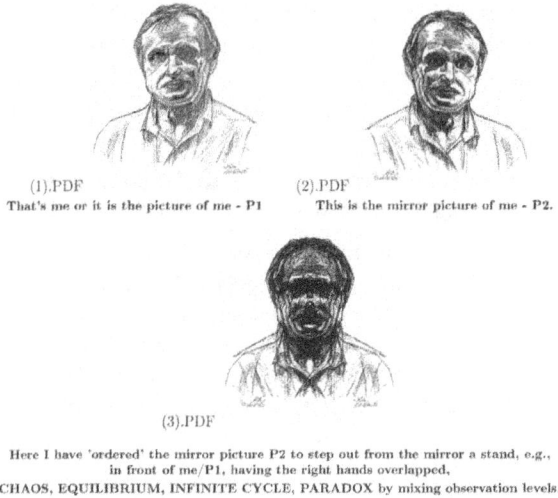

(1).PDF
That's me or it is the picture of me - P1

(2).PDF
This is the mirror picture of me - P2.

(3).PDF

Here I have 'ordered' the mirror picture P2 to step out from the mirror a stand, e.g.,
in front of me/P1, having the right hands overlapped,
CHAOS, EQUILIBRIUM, INFINITE CYCLE, PARADOX by mixing observation levels.

Figure A1.
Example of not distinguishing the reality and its image.

A.2 The proof way of Caratheodory theorems

I. Let the form $\delta Q = \sum_{i=1}^{n} X_i x_i$ has the integration factor v and let
$d\mathcal{R} = \sum_{i=1}^{n} \frac{1}{v} X_i dx_i$. Then the Pfaff equation $\delta Q = \sum_{i=1}^{n} X_i dx_i = 0$ has the solution in
the form $\mathcal{R}(x_1, \ldots, x_k) = $ const. and this solution represents a family of hyperplanes
in n-dimensional space, not intersecting each other. **Let us pick now the point**
$P\left(x_1^0, \ldots, x_n^0\right)$ **determined by our choice of** const. = C. **Only the points lying in the**
hyperplane $\mathcal{R}\left(x_1^0, \ldots, x_n^0\right)$ **are accessible from the point** P **along the path satisfy-**
ing the condition dQ = 0. **All the points not lying in this hyperplane are inaccessi-**
ble from the point P **along the path satisfying the condition** dQ = 0 **(Figure A2).**

II. Let us pick the point V, e.g., from $\mathbb{R}^3$, lying in a vicinity of the point P, which
is not accessible from P following the path dQ = 0. Let g be a line going through the
point P and let g be oriented $(\vec{g})$ in such way that it does not satisfy the condition
dQ = 0. The point V and the line g determine a plane $X_i = X_i(u, v)$, i = 1, 2, 3. Let us

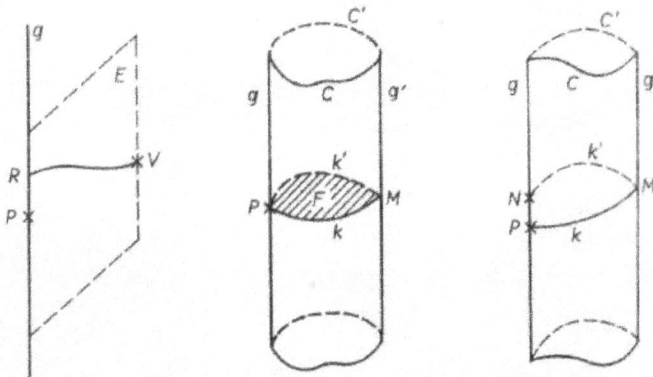

Figure A2.
The proof way of the Caratheodory theorems.

consider a curve k in this plane, going through the point $V(u_0, v_0)$ in that way $(\vec{g})$ that dQ = 0 is supposedly valid along this curve. **There is only one curve k for the point $V(u_0, v_0)$. It lies in our plane, the plane $X_i = X_i(u, v)$, and then it is valid for it d$X_i = \frac{\partial X_i}{\partial u} du + \frac{\partial X_i}{\partial v} dv$ and, considering dQ = 0 along k, we get $\sum_{i=1}^{3} X_i \frac{\partial X_i}{\partial u} du +$ $\sum_{i=1}^{3} X_i \frac{\partial X_i}{\partial v} dv = 0$.**

The curve k, however, intersects the line g in the point R, which is inaccessible from the point P along the path with dQ = 0 (for $dQ_{R_{\vec{g}}} \neq 0$). Otherwise, the point V would also be accessible from the point P through R and k $\left(dQ_{R_k} = 0\right)$, which is a conflict with the original assumption. By a suitable selection of V, it is possible to have the point R arbitrarily close to the point P; in the arbitrary vicinity of the point P, there are points inaccessible from the point P along the path with dQ = 0. Now, let us pick a line g' parallel to the line g, and a cylinder C going through these two lines. We consider that the curve k satisfying the relation dQ = 0 is on this cylinder C' goes through the point P and intersects the line g' in the point M.

Now, let us consider another cylinder C' as the continuation of C with g' and g. Let us use the symbol k' for the continuation of the curve k in C'. Then the curve k' must intersect the line g in the point P. Otherwise, it would be possible to deform the plane C' as much as to get C, thus continually merging the intersecting point N into the point P and at the moments of discrepancy of the points P and N, it would be possible to reach the point P from the point N along the line g (supposedly with dQ = 0). However, the condition dQ = 0 is not valid there $(dQ_{R_{\vec{g}}} \neq 0)$. By deforming C' into C, the k and k' would close a plane F where dQ = 0. If the equation of this plane has the form $\mathcal{R}(x_i)_{i=1}^{3} = \text{const.}$, then the equation dQ = 0 has a solution—an integration factor for the **Pfaff form** $\delta Q = \sum_{i=1}^{3} X_i dx_i$ exists [11].

A.3 Information thermodynamic concept removing autoreference

The concept for ceasing the autoreference, based on the two Carnot Cycles disconnected as for their heaters and described informationally, shows the following **Figure A3**. (also see [1, 2, 4]):

For ΔA, it is valid in the cycle $\mathcal{O}''$ that

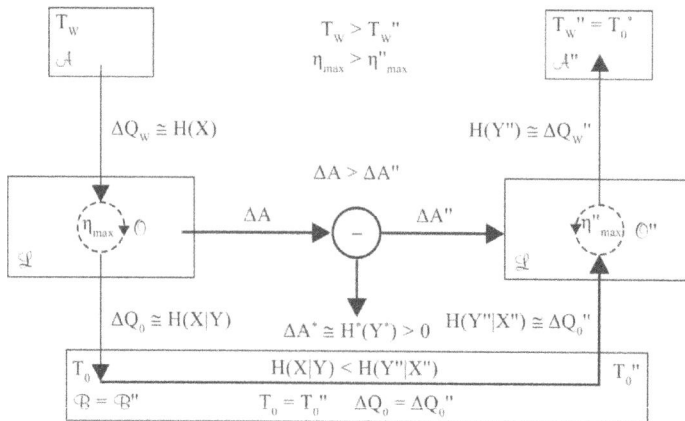

Figure A3.
The concept for ceasing the autoreference.

$$\Delta A'' = \Delta Q_W'' \cdot \left(1 - \frac{T_0}{T_W''}\right) = \Delta Q_W \cdot \frac{T_W''}{T_W} \cdot \left(1 - \frac{T_0}{T_W''}\right) =$$

$$= \Delta Q_W \cdot \left(\frac{T_W''}{T_W} - \frac{T_0}{T_W}\right) = k \cdot H(X) \cdot (T_W'' - T_0) \qquad (26)$$

$$= k \cdot H(X) \cdot T_W'' \left(1 - \frac{T_0}{T_W''}\right) = k \cdot H(X) \cdot T_W'' (1 - \beta'') = k \cdot T_W'' \cdot H(Y'')$$

and, further, for ΔA in the cycle $\mathcal{O}$, we have

$$\Delta A = k \cdot H(X) \cdot T_W (1 - \beta) = k \cdot H(X) \cdot T_W \left(1 - \frac{T_0}{T_W}\right) \qquad (27)$$

and thus, for the cycles $\mathcal{O}''$ and $\mathcal{O}$, it is valid that

$$\frac{\Delta A''}{k T_W''} = H(X) \cdot \left(1 - \frac{T_0}{T_W''}\right) = H(X) \cdot (1 - \beta'') = H(X) \cdot \eta_{max}''$$

$$\frac{\Delta A}{k T_W} = H(X) \cdot \left(1 - \frac{T_0}{T_W}\right) = H(X) \cdot (1 - \beta) = H(X) \cdot \eta_{max} \qquad (28)$$

For the whole work ΔA^* of the combined cycle $\mathcal{O}\mathcal{O}''$, we have

$$\Delta A^* = \Delta A - \Delta A'' = \left[k T_W \cdot H(X) \cdot (1 - \beta) - k T_W'' \cdot H(X) \cdot (1 - \beta'')\right] > 0 \qquad (29)$$

Then, for the whole change of the thermodynamic entropy within the combined cycle $\mathcal{O}\mathcal{O}''$ (measured in information units *Hartley, nat, bit*) and thus for the change of the whole information entropy $H^*(Y^*)$, it is valid that

$$H^*(Y^*) = \frac{\Delta A^*}{k T_W} = H(X) \cdot \left[(1 - \beta) - \frac{T_W''}{T_W} \cdot (1 - \beta')\right]$$

$$= H(X) \cdot \left(1 - \frac{T_0}{T_W} - \frac{T_W''}{T_W} + \frac{T_0}{T_W}\right) = H(X) \cdot \left(1 - \frac{T_W''}{T_W}\right) \qquad (30)$$

It is valid, for ΔA^* is a *residuum work* after the work ΔA has been performed at the temperature T_W. Evidently, the sense of the symbol T_W'' (within the double cycle $\mathcal{O}\mathcal{O}''$ and when $\Delta Q_0 = \Delta Q_0$) is expressible by the symbol T^*_0, which is possible, for the working temperatures of the whole cycle $\mathcal{O}\mathcal{O}''$ are T_W and $T_W = T^*_0$. The relation (30) expresses that fact that the double cycle $\mathcal{O}\mathcal{O}''$ is the direct Carnot Cycle just with its working temperatures $T_W > T_W = T^*_0$. In the double cycle $\mathcal{O}\mathcal{O}''$, it is valid that

$$\beta'' = \frac{\Delta Q_0''}{\Delta Q_W''} = \frac{\dfrac{\Delta Q_0''}{T_W''}}{\dfrac{\Delta Q_W''}{T_W''}} = \frac{H(Y''|X'')}{H(Y'')} = \frac{T_0}{T_W''}, T_W'' = T^*_0, \text{ cyklus } \mathcal{O}''$$

$$\beta = \frac{\Delta Q_0}{\Delta Q_W} = \frac{\dfrac{\Delta Q_0}{T_W}}{\dfrac{\Delta Q_W}{T_W}} = \frac{H(X|Y)}{H(X)} = \frac{T_0}{T_W}, \text{ cyklus } \mathcal{O} \qquad (31)$$

$$\frac{\beta}{\beta''} = \frac{T_W''}{T_W} = \frac{T^*_0}{T_W} \triangleq \beta^*$$

and then, by (30) and (31) is writable that

$$\frac{\Delta A^*}{kT_W} = H(X) \cdot (1 - \beta^*) = H(X) \cdot \left[1 - \frac{H(X|Y) \cdot H(Y'')}{H(Y''|X'') \cdot H(X)} \right] > 0 \tag{32}$$

It is ensured by the propositions $T_W > T_W, T_0 = T_0$ and also by that fact that the loss entropy $H(X|Y)$ is described and given by the heat $\Delta Q_0 = \Delta Q'_0$. But in our combined cycle $\mathcal{OO''}$, it is valid too that

$$H(X) = \frac{\Delta Q_W}{kT_W} = \frac{\Delta Q_W''}{kT_W''} = H(Y'') \quad \left[= \frac{\Delta Q_W''}{kT^*_0} \right] \tag{33}$$

and we have

$$\frac{H(X|Y)}{H(Y''|X'')} = \beta^* < 1 \tag{34}$$

For the whole information entropy $\frac{\Delta A^*}{kT_W}$ (the whole thermodynamic entropy $\mathcal{S}_C$ in information units) and by following the previous relations also it is valid that

$$\frac{\Delta A^*}{kT_W} = H(Y'') - H(Y'') \cdot \beta^* = H(Y'') \cdot \left(1 - \frac{T^*_0}{T_W} \right)$$
$$= H(Y'') \cdot \left[1 - \frac{H(X|Y)}{H(X''|Y'')} \right] \tag{35}$$

And thus, the structure of the information transfer channel $\mathcal{K}$ [expressed by the quantity $H(X|Y)$] is measurable by the value $H^*(Y^*)$ from (32) and (35). Symbolically, we can write, using a certain growing function f,

$$H^*(Y^*) = \frac{\Delta A^*}{kT_W} \cong f[H(X|Y)] > 0 \tag{36}$$

The cycles $\mathcal{O}, \mathcal{O''}$, and $\mathcal{OO''}$ are the Carnot Cycles, and thus from their definition and construction, they are imaginatively[16] in principle, the infinite cycles; in each of them the following *criterion of an infinite cycle* (see [12]) it is valid inevitably,

$$T\left(X^{[\cdot]}; Y^{[\cdot]} \right) = H\left(X^{[\cdot]} \right) - H\left(X^{[\cdot]} | Y^{[\cdot]} \right) = H\left(Y^{[\cdot]} \right) > 0 \text{ and } \Delta \mathcal{S}^{[\cdot]}_{\mathcal{L}} = 0 \tag{37}$$

The construction of the cycle $\mathcal{OO''}$ enables us to recognize that the infinite cycle $\mathcal{O}$ is running. In our case, it is the infinite cycle from (5), (6) and also from [4, 8, 10],

$$\begin{aligned}
&Q(\mathbb{X}, \ Y), \quad Q[\mathbb{X}, \ \Phi[Q(\mathbb{X}, \ Y)]], \quad Q[\mathbb{X}, \ \Phi[Q(\mathbb{X}, \ \Phi[Q(\mathbb{X}, \ Y)])]], \ \ldots \\
&Q(\mathbb{N}_0, \ Y), \quad Q[\mathbb{N}_0, \Phi[Q(\mathbb{N}_0, \ Y)]], \quad Q[\mathbb{N}_0, \Phi[Q(\mathbb{N}_0, \ \Phi[Q(\mathbb{N}_0, \ Y)])]], \ \ldots
\end{aligned} \tag{38}$$

---

[16] When an infinite reserve of energy would exist.

Author details

Bohdan Hejna
Department of Mathematics, University of Chemistry and Technology, Prague,
Czech Republic

*Address all correspondence to: hejnab@vscht.cz

IntechOpen

References

[1] Hejna B. Recognizing the infinite cycle: A way of looking at the halting problem. In: Dubois DM, editor. Proceedings of the Tenth International Conference CASYS'11 on Computing Anticipatory Systems, Lecture on CASYS'11 Conference, 8–13 August 2011. CHAOS; 2012. ISSN: 1373-5411

[2] Hejna B. Informační termodynamika III.: Automaty, termodynamika, přenos informace, výpočet a problém, zastavení. Praha, VŠCHT Praha; 2013. ISBN: 978-80-7080-851-1

[3] Hejna B. Information thermodynamics and halting problem. In: Bandpy MG, editor. Recent Advances in Thermo and Fluid Dynamics. Croatia, Rijeka: InTech; 2015. pp. 127-172. ISBN: 978-953-51-2239-5. Available from:http://www.intechopen.com/books/recent-advances-in-thermo-and-fluid-dynamics

[4] Hejna B. Gödel proof, information transfer and thermodynamics. In: Lecture on IIAS Conference; 3–8 August 2015; Baden-Baden, Germany; Journal IIAS-Transactions on Systems Research and Cybernetics. The Inernational Institute for Advanced Studies in System Research and Cybernetics; 2015; **15**(2). IBSN: 978-897546-13-0

[5] Gödel K. Über formal unentscheidebare Satze der Principia Mathematica und verwandter Systeme I. von Kurt Godel in Wien; Monatshefte fur Mathematik und Physik 1931;38:173-198

[6] Gödel K. On Formally Undecidable Proposition of Principia Mathematica and Related Systems. Vienna; 1931 (translated by B. Metzer)

[7] Včelař F, Frýdek J, Zelinka I. Godel 1931. Praha: Nakladetelství BEN; 2009

[8] Hejna B. Information transfer and thermodynamics point of view on Gödel proof. In: Thomas C, editor. Ontology in Information Science; College of Engineering Trivandrum, India. InTech; 2017/18. ISBN: 978-953-51-5354-2, ISBN: 978-953-513888-4. Print ISBN: 978-953-51-3888-4. Available from: http://www.intechopen.com/books/ontology-in-information-seience; OAI link: http://www.intechopen.com/oai/?verb=ListIdentifiersmetadataPrefix=oai$_d$cset=978-953–51–3887-7; Scientometrics on: https://www.intechopen.com/books/statistics/ontology-in-information-science/information-transfer-and-thermodynamic-point-of-view-on-gödel-proof; Indexing: WorldCat, Base, A to Z, IET Inspect, Scirus, Google Scholar

[9] Hejna B. Informační termodynamika IV.: Gödelovy věty, přenos informace, termodynamika a Caratheodoryho věty. Praha: VŠCHT Praha; 2017. ISBN: 978-80-7080-985-3

[10] Hejna B. Gödel and Caratheodory theorems. In: Lecture on IIAS Conference; August 2017; Baden-Baden, Germany. Journal IIAS-Transactions on Systems Research and Cybernetics. 2017;1. ISSN:1609-8625. ISBN: 978-1897456-42-0

[11] Hála E. Úvod do chemické termodynamiky. Praha: Academia; 1975

[12] Hejna B. Information thermodynamics. In: Moreno-Piraján JC, editor. Thermodynamics—Physical Chemistry of Aqueous Systems. Croatia, Rijeka: InTech; 2011. pp. 73-104. ISBN: 978-953-307-979-0. Available from: http://www.intechopen.com/articles/show/title/information-thermodynamics

Chapter 2

Knowledge Patterns within the Conception of Semantic Web

Martin Žáček, Alena Lukasová, Marek Vajgl and Petr Raunigr

Abstract

The article tries to contribute to answer a question if the general concept of knowledge pattern with its sub-concepts covers a great majority of the approaches used under this term in computer science literature. At one case, specialized software design patterns in the frame of object-oriented methodology become a very well used tool for software praxis; at a different case, there exists a large packet of tools for creating ontologies of various areas. As a third case, also RDF-based networks of linked data could be seen as knowledge patterns characterizing at least structures or defined activities of some social, working, or other organizations. We propose here to see the problem of knowledge pattern from knowledge representation especially at directions where the goal of using knowledge pattern meets the general goal of the semantic web. The motivation of this article is to apply knowledge patterns in the semantic web because knowledge at a higher professional level can and should usually be given in such a way that their specialized formal expertise incorporates the key to understanding their meaning.

Keywords: pattern, knowledge, RDF, CFL, semantic web

1. Introduction

Generally, the concept of knowledge pattern [1] modeling appears in knowledge engineering, apparently due to the corresponding concept of a design pattern in software development in the frame of object-oriented methodology. But the difference is mainly in the area of the two points of view. While the design pattern is focused toward general principles of software creation in terms of practices, structure, or behavior properties, the corresponding specifications of knowledge patterns need to take into focus minimally the concept of knowledge, its properties, and cases. While in the case of declarative knowledge it simply involves acquiring new knowledge or its new application from a given knowledge base, all within the first-order logics formalism, for procedural knowledge a generally acceptable formal language and approach until now has not been found. But knowledge pattern for procedural knowledge case gives us a possibility to use similar rules of design patterns as well as in the case of software development. If moreover knowledge pattern has been embedded into the semantic web concept [2] environment with a seeing the world throw the RDF principle [3], it represents a new quality in the sense that content and form become easy-to-use for computers and comprehensible to users without deeper penetration into the principles of knowledge engineering.

2. How to take the topic of knowledge into the AI?

At the Cambridge English Dictionary, we can read a definition of the meaning of the concept of knowledge as follows:

> *An understanding of or information about a subject that you get by experience or study, either known by one person or by people generally.*

The question in the title of the paragraph with a corresponding Cambridge explanation seems to be the basic one. But our goal must be a bit more different from a topic at philosophy (gnoseology); we only try to generalize a bit an orientation about the concept of knowledge pattern within a formal representation language used in AI.

2.1 What is knowledge in praxis about?

In the epistemological area of artificial intelligence, we encounter formal manipulation of knowledge. Knowledge is based on information, and this information is based on data [4].

To be able to work with knowledge in a form suitable for computer implementation, it is necessary to introduce some formalism—a representative language. This language must be able to reflect the relationship between knowledge of the world, stored in human minds, and knowledge written in formal means.

Knowledge is information that is usable and divisible, respectively, in relation to other information [4].

In other words:

- In one case, it is about what entities a specified (reference) "world" consists of and what are their properties and relationships. The description of the status of participating entities has a declarative character in this case.

- In the latter case, it is about the starting state of the "world" and using rules for reaching a target state. The description here is of a procedural form, capturing the crucial interstates through which the process passes.

2.2 Knowledge elements and knowledge components

Knowledge elements are bounded to a certain knowledge base written usually in a special formal approach (or language syntax). The elements are atoms that cannot be further divided. An important feature of atoms is their independence from external contexts. This is especially important for their applications and reuse. Knowledge must be lasting about the knowledge base in which knowledge can be manipulated; it must have a permanent meaning.

Knowing elements linked to a given knowledge base can be composed into knowledge components, whose syntax and meaning is created through the grammatical rules and (logical) composition of participating atoms [5].

Knowledge components are conceptually dependent on the knowledge model used and on its required properties. The factors that make up the overall character of the knowledge component in the composition should be the following:

- The functionality is reflecting and sharing a specified relationship between the start state and target state.

- Reliability, which means that the component is mature.

- Applicability and portability are that the component is understandable and appropriately applicable to allow wider use.

- Modifiability, i.e., the ease of partial changes in the stability of the basic properties.

The meaning of the elementary or compound knowledge is secured at all points above if the RDF model has been chosen to represent a conceptual reality.

2.3 What is a knowledge pattern?

The term "knowledge pattern" was first used in [1, 6]. While building ontologies or knowledge bases, one can see that some structures of modeled knowledge are the same. These same structures of knowledge can be captured as knowledge patterns. Knowledge patterns are general structures (patterns) of knowledge, which are not a part of the target knowledge base. They can be included into a target knowledge base by renaming their nonlogical symbols. This renaming is called morphism. The morphism is an important part of using knowledge patterns [6, 7].

Presently, there is no direction for capturing knowledge patterns. We propose to model knowledge patterns in RDF graph models [8, 9] of the semantic web.

Going through the topics of knowledge pattern at the web, we have to meet the following more or less similar approaches:

- Seeking general knowledge pattern as (1) a small ontology [10] within a well-specified part of the (reference) world.

- Seeking general knowledge pattern as (2) a frame structure of linked data of some oft appearing kernels of together-linked facts about everyday life, as is the case of a firm leading structure and tasks of participants.

- Seeking general knowledge as (3) the helping means on how to construct a special product or software of expected properties.

The first and second cases represent more or less declarative approaches to reach the main result from them. They are of hierarchical strictures of classes with their subclasses. The third one represents a procedural knowledge, but it is difficult to find a common principle of building results as a formal description. But in this case, there is a rich database of very useful prescribes for a big scale in praxis.

A specification of the concept in the title within the formal representation approach can be something like the following:

A knowledge pattern [6] concerns "holding as true of a set of sentences or rules about a specified piece of the world—either known by one person or by people generally, all expressed by formal means."

2.4 Knowledge pattern (KP) within a knowledge representation

In general, in the field of types of knowledge, two basic cases corresponding to the kinds of knowledge at Section 2.1 of the access in formalization should be considered:

1. Declarative knowledge representation

2. Procedural knowledge representation

In summary, a common feature in both types of pattern specification is their generic validity within a given environment, verification by historical development, and long-term experience in their applications. If moreover knowledge design patterns are embedded in the semantic web concept environment, they represent a new quality in the sense that their content and form become comprehensible to users without deeper penetration into the principles of knowledge engineering.

In terms of global formal representation, the question is how to create:

3. A unifying approach is generally applicable to the representation of knowledge (sometimes also appearing under the working name "framework approach") based on both of the above approaches, enriched by those means of representation that are missing in the first or second access.

We would like to present here a proposal to use instead of the ontology design patterns [11] for the approach to the topic a simple knowledge pattern bounded especially to specialized kind of knowledge.

2.5 Knowledge pattern (KP) versus special design patterns

A majority of authors use the term ontology design pattern (OP) because the OP is, in fact, a modeling solution of solving a recurrent ontology design problem. We would like to show here the fact that the attribute "ontology" is not a necessity in the case of using RDF modeling principle that carries this property as an implicit one.

Definition of the OP according to the authors of the article [11]:

Ontology design pattern (or only OP) is a modeling solution to solve a recurrent ontology design problem.

Authors [12] have identified within web documents several types of OPs and have grouped them into six families:

- Structural OPs

- Correspondence OPs

- Content OPs (CPs)

- Reasoning OPs

- Presentation OPs

- Lexico-syntactic OPs

Knowledge patterns [13] of declarative or procedural knowledge cases give a possibility how to use rules of similar design patterns as well as in the case of software development. In both cases a natural language plays an important role, and definition in a special language is not necessary.

While examples of software design pattern applications can be found to a large extent, knowledge patterns tend to be related only in a few types of problem areas and their language representations.

3. Knowledge in formal representation

3.1 Representation of declarative knowledge

At the beginning, the principle of seeing a simplified world was considered an abstraction of the real world to be modeled. Usually, the E-R principal is to be chosen.

The process of conceptualization has stabilized during the development of the means of modeling reality onto the well-established world view as a set of entities with certain characteristics and mutual relationships.

Now at first a generally recognized and application-proven way of using a formal language is given for the sake of building a formal description of concepts and their properties based on the basic (conceptual) level of the E-R model world abstraction [14]. Conceptualization is partially subordinated to the expected formal language syntax. The semantics of the language of knowledge represented in this way of seeing the world ought to be derived from declarative descriptions of the properties and relationships of the entities of the given reference world. They are then formally represented according to the rules applicable in the reference world. From anchoring corresponding concepts on the web or the semantic web, the current state of development has to choose the use of the principle of RDF modeling [15]. Using the RDF [15] data model representation gives a possibility of graphic representation of RDF triples [16] as vectors expressing corresponding knowledge elements (see **Figure 1**): <subject><property><object>.

RDF describes the resource (as a subject), which has some property with a corresponding value (object). The RDF model is based on associative (semantic) networks [17, 18].

In the following figure (**Figure 1**) we can see the conversion of the sentence "Marek teaches the subject of pgm languages." Clearly here we see that the subject is "Marek," the object is "pgm language," and the property is "teaches."

3.2 Representation of procedural knowledge

The most successfully applied approach to the representation of procedural properties of the modeled world is the output of the process of algorithmic representation of the modeled reality, which builds a formal description of the reference world based on the graphic expression of the formalization—a flowchart of the basic elements of human activity in it.

Just as a formalization of declarative knowledge is guided by a conceptual flowchart, procedural knowledge [19] is the guiding factor of the problem of its algorithm, i.e., the way of seeing the process described based on elementary programming language components with special language syntax. The

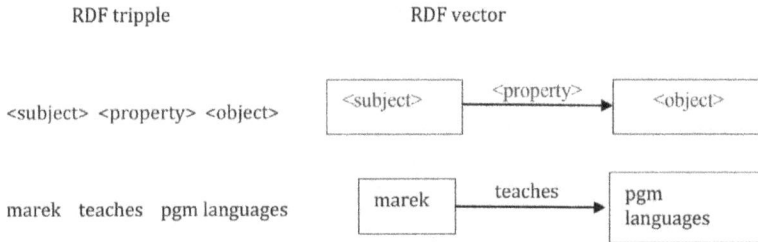

Figure 1.
Representation of the fact "Marek teaches the subject of pgm languages" using a RDF triple and with the help of RDF graph vectors.

representative language used to simulate the performance of the modeled activity usually has a specific form of a programming language reflecting the characteristics of the modeled activity and the practical application needs.

Algorithms typically ignore entities with their properties and relationships, with the interconnection of modeling activities in the process providing "data," usually without any closer anchoring in entity representations. The problem solves the following frame approach of modeling.

3.3 The framework character of knowledge

Algorithmic representations of the modeled world with its procedural properties, as well as in the case of point 1, must take into a focus relation to elementary entities and their properties and relationships of a modeled world. It should, therefore, use a representation method based on the RDF model corresponding to point 1, with the terminology relating to entities with their properties and relationships being anchored in the chosen dictionary (ontology) to which access created by the RDF model [20] has been bound.

E.g. the language UML (as a means of describing RDF-modeled reality) would allow UML diagrams to data retrieve into the represented process and their belonging to home entities within the chosen ontology.

4. Knowledge pattern in a semantic web context

4.1 RDF modeling principle and knowledge patterns

Creating data for the semantic web means conceptualizing world using E-R model with a participation of a key ontology because of sharing and reusing formalized knowledge representation. Each data item can take its meaning from a standardized description of web resources within the proper ontology using its URI identifier.

4.2 Semantic web patterns and anti-patterns

The RDF as a general framework for describing, replacing, and reusing metadata represents the technological foundation of the semantic web. From anchoring corresponding concepts on the web (or the semantic web), the current state of development is the use of the RDF [4, 21] modeling principle.

RDF describes the resource, which has some properties, and these properties have corresponding values (**Figure 1**). While the subject defines the source, the property determines its nature and at the same time expresses the relationship between the subject and the object.

The semantic web idea is based on the RDF technology [22], which integrates the web language syntax and the naming of its elements by URIs. So a content presented on the semantic web has a well-defined meaning and allows a better understanding of both people and software agents.

The semantic web provides a common framework that allows data to be shared and reused.

It also emphasizes the ease of understanding and applicability of documents on the web, especially easy usability of knowledge model as well as knowledge pattern approach.

4.2.1 Definition

The knowledge pattern is a general type of component knowledge of proven success, often with a design concept of good practice, a process of structuring, to create the architecture of component knowledge. It may be declarative, procedural, or frame-like.

An anti-pattern is a common often accompanying process phenomenon that is not involved in solving the problem (wrong solutions or "worst-case" solutions). Unlike the model, the anti-model generally describes individual non-model cases and highlights a general solution to recurring problems.

In its formal representation, the knowledge pattern should:

- Be anchored within a specific knowledge base with a given semantics.

- If possible be used for the formal representation of knowledge through a language with easy-to-understand interpretations (preferably graphical).

- Have individual atomic components, of which the pattern/anti-pattern is composed, clearly defined, and described.

4.3 An example of a knowledge pattern for the case of declarative knowledge

As an example of the knowledge pattern, we shall show in this paragraph the case of a declarative knowledge at its basic logical form.

In the field of formal logic [23], the declarative knowledge design pattern extends deep into the history of formal systems. At a time when philosophers and mathematicians changed their orientation from specific individual descriptions to general principles of reasoning, they came as a result of general principles of deduction in formal logics [23]. Procedures as a rule modus ponens or the resolution inference rule in propositional or predicate logic represent in terms of the ongoing development of artificial intelligence typically general guidance on how to derive from the assumption's logical consequences, respectively how to arrive at a logical deduction on the arguments that confirm or reject given assertions. This is nothing more than a guideline on the application of a knowledge pattern.

5. Rules of deduction in formal logics in the role of declarative knowledge patterns

The publication on formal logic and semantic web [4] lists several examples of the application of a resolution deduction rule as a knowledge model allowing the derivation of a logical consequence from given assumptions.

The following example illustrates the use of RDF CFL resolution derivation rule that obtained a logical consequent from a knowledge base [4].

An example of immigration rules for Europe is given as a knowledge base in the language of the first-order predicate logic, which, as well known, is not one of the easy-to-understand and usable languages of formal logic. However, there is a way of transferring (according to well-known rules) to the special clause of CFL [24], which has been before based on conceptualization according to the RDF principle. To express its concepts and their properties and relationships, this language, on behalf of RDF CFL [25], uses exclusively the binary predicates. Consequently, a corresponding graphical representation has been used, playing an important role in the semantic web.

5.1 Example

Immigration rules for citizen as a knowledge base in the FOPL

1. $\forall x \, \forall y$ (**stateEU**(y) & **citizen**(x,y) → **enter**(x))

2. $\forall x \, \forall y$ (¬**stateEU**(y) & **citizen**(x,y) & **has_visa**(x) → **enter**(x))

citizen(anne, aus), **citizen**(achim,tur), **has-visa**(achim),
stateEU(aus), ¬**stateEU**(tur).

5.1.1 Resolution rule in a generalized form

Creating a CFL clause for using the resolution rule of the RDF CLF language according to the scheme:

$$\frac{< \text{CFL clause 1} > \qquad\qquad <\text{CFL clause 2} >}{< \text{the logical consequent of 1 and 2} >}$$

As the set of the RDF CFL clausal form contains only positive atoms, we prepare for the resolution rule those basic atoms that express positive statements about the participating persons and states in our example (**Table 1**):

1. **stateEU**(aus), **citizen**(anne, aus) → **enter**(anne).

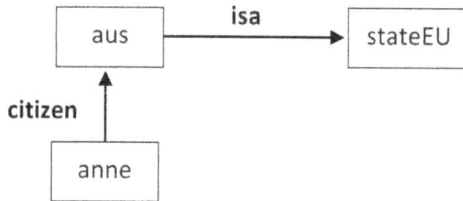

Verbal expression: "Anne is a **citizen** of Austria, which **is a** state of EU."

2. RDF CFL instance of the (1.)

    → **stateEU**(aus).

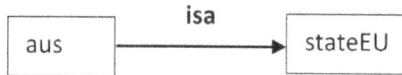

Verbal expression: "Austria **is a** state of EU."
citizen(anne, aus) → **enter**(anne).
Logical consequent after a using of the resolution rule.

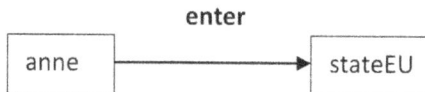

Verbal expression: "Anne can enter to any state of EU."
Similar examples can be seen in the publication [26].

Entity	IRI (URI)
state	http://dbpedia.org/page/State_(polity)
citizen	http://dbpedia.org/page/Citizenship
aus	https://www.wikidata.org/wiki/Q40
anne	http://dbpedia.org/page/Anne
visa	http://dbpedia.org/page/Travel_visa
enter	https://www.wikidata.org/wiki/Q161935
isa	is a

Table 1.
IRI of the example (RDF).

6. A proposed structure of records in the knowledge pattern archive

Design Name	A unique name that adequately describes the knowledge pattern and reference it helps identify.
Objective	A description of the goal for which the knowledge pattern originated.
Also known as	Other names for the same knowledge pattern.
Context	A situation where a description of the knowledge pattern is useful.
Motivation⋆	Description of the problem to be solved by the given knowledge model.
Usage	Situations in which the pattern can be used. This is the context of a knowledge model.
Structure	Graphic representation of the knowledge pattern.
Participants	List of classes and objects that use this knowledge pattern and their role in the design.
Consequences	A description of the results, side effects, and problems that the pattern makes use of.
Known uses	Examples of practical use of the pattern.

⋆contradictory forces.

7. The RDF CFL graph resolution rule in the proposed pattern archive

Design Name	RDF CFL graph resolution rule.
Objective	The pattern in the form of a rule serves a possibility of obtaining further knowledge like consequents of a knowledge base.
Also known as	-
Context	Some of the consequents hidden before become visible and could make the whole content more understandable.
Motivation⋆	Better level of usability of a knowledge base.
Usage	as an example: solving the consequent from two clauses in the graph representation.
Structure	Symbolic expression of a deduction.

    <graph of clause 1>   <graph of clause 2>

        <graph of the logical consequent>

Participants	Graph records of two clauses with at least one identical atom.
Consequences	If an empty clause has been obtained, it means an unsolvability of the input clauses without any consequent.
Known uses	An example of practical use of the pattern seen above at example 1.

*contradictory forces.

8. Conclusion

The idea of a semantic web can be a handy hand given to those who still feel at least a mistrust, if not resistance, to take knowledge from websites. Knowledge from areas at a high level of science can be discouraged by a layman, among other things, due to ignorance of the special language of the field in question. Thus, any known common patterns, which can be called "knowledge patterns," are not recognized in the new context, so they are generally not considered to be reusable. It is, therefore, necessary to look at approaches that can convey wider usability. The semantic web has this approach in the very description of its definition. Knowledge at a higher professional level can and should usually be given in such a way that their specialized formal expertise incorporates the key to understanding their meaning. Our goal is not to create a universal tutorial, but we must integrate the semantics of knowledge presented in the formal language direct into its syntax. The above graphic example is illustrative.

Author details

Martin Žáček*, Alena Lukasová, Marek Vajgl and Petr Raunigr
Faculty of Science, University of Ostrava, Ostrava, Czech Republic

*Address all correspondence to: martin.zacek@osu.cz

References

[1] Clark P, Thompson J, Porter B. Knowledge patterns. In: Staab S, Studer R, editors. Handbook on Ontologies. Berlin: Springer-Verlag; 2004. pp. 191-207. ISBN: 3-540-40834-7

[2] Berners-Lee T, Hendler J, Lassila O. The semantic web. Scientific American. 2001;**284**(5):28-37

[3] Swartz A. The Semantic Web in Breadth [Internet]. 2002. Available from: http://logicerror.com/semantic Web-long

[4] Lukasová A, Žáček M, Vajgl M, Telnarová Z. Formal Logics and Semantic Web (Formální Logika a Sémantický Web). Plzeň: ZČU Plzeň; 2015. ISBN: 978-80-261-0408-7

[5] Schwitter R. Controlled natural languages for knowledge representation. In: Proceedings of the 23rd International Conference on Computational Linguistics; 2010. pp. 1113-1121

[6] Miarka R, Žáček M. Knowledge patterns for conversion of sentences in natural language into RDF graph language. In: Proceedings of the Federated Conference on Computer Science and Information Systems. USA: IEEE Computer Society Press; 2011. pp. 63-68. ISBN: 978-1-4577-0041-5

[7] Miarka R, Žáček M. Knowledge patterns in RDF graph language for English sentences. In: Proceedings of the Federated Conference on Computer Science and Information Systems. USA: IEEE; 2012. pp. 109-115. ISBN: 978-83-60810-48-4

[8] W3C. Resource Description Framework (RDF): Concepts and Abstract Syntax [Internet]. Available from: http://www.w3.org/TR/2004/REC-rdf-concepts-20040210/

[9] W3C. RDF Primer [Internet]. Available from: http://www.w3.org/TR/2004/REC-rdf-primer-20040210/

[10] Žáček M. Ontology or formal ontology. In: International Conference of Numerical Analysis and Applied Mathematics 2016 (ICNAAM 2016): AIP Conference Proceedings. American Institute of Physics Inc.; 2017. ISBN: 978-073541538-6

[11] Gangemi A. Ontology design patterns for semantic web content. In: Gil Y, et al. editors. ISWC 2005; 2005. pp. 262-276

[12] Rech J, Feldmann R, Fraunhofe E. Knowledge Patterns [Internet]. Available from: http://joerg-rech.com/Paper/RechFeldmannRas_EncycKM_KnowledgePatterns.pdf

[13] Rech J, Decker B, Jedlitschka A. The quality of knowledge: Knowledge patterns and knowledge refactorings source. International Journal of Knowledge Management (IJKM). 2007. pp. 578-586

[14] Žáček M, Lukasová A, Vajgl M. Ontology languages for semantic web from a bit higher level of generality. In: 10th International Scientific Conferences on Research and Applications in the Field of Intelligent Information and Database Systems; ACIIDS 2018: Lecture Notes in Computer Science 2018-03-19, Dong Hoi City, Vietnam. Switzerland: Springer Verlag; 2018. pp. 275-284. ISBN: 978-331975419-2

[15] W3C. Resource Description Framework (RDF) Model and Syntax Specification [Internet]. 2008. Available from: http://www.w3.org/TR/1999/REC-rdf-syntax-19990222

[16] Lukasová A, Žáček M, Vajgl M. Reasoning in formal systems of

extended RDF networks. In: 9th Asian Conference on Intelligent Information and Database Systems (ACIIDS): Intelligent Information and Database Systems, Kanazawa, Japan. Switzerland: Springer Verlag; 2017. pp. 371-381. ISBN: 978-3-319-54430-4

[17] Findler NV, editor. Associative Networks: Representation and Use of Knowledge by Computers. New York: Academic Press; 2004

[18] Lukasová A, Vajgl M. Žáček M. Knowledge represented using RDF semantic network in the concept of semantic web. In: Proceedings of the International Conference on Numerical Analysis and Applied Mathematics 2015 (ICNAAM-2015). USA: American Institute of Physics Inc.; 2016. ISBN: 978-0-7354-1392-4

[19] Vajgl M, Lukasová A. Žáček M. Knowledge bases built on web languages from the point of view of predicate logics. In: Proceedings of the International Conference on Applied Mathematics and Computer Science, ICAMCS 2017: AIP Conference. American Institute of Physics Inc.; 2017. ISBN: 978-073541506-5

[20] Lukasová A, Žáček M. Creating words by inflexion and derivation in RDFCFL graphs. WSEAS Transactions on Computer Research. 2016;4(23): 208-212. ISSN: 1991-8755

[21] Staab S, Erdmann M, Maedche A, Decker S. An extensible approach for modeling ontologies in RDF (S). In: Knowledge Media in Healthcare: Opportunities and Challenges. IGI Global; 2002. pp. 234-253. https://www.igi-global.com/chapter/knowledge-media-healthcare/25416

[22] Gangemi A. Ontology design patterns for semantic web content. In: International Semantic Web Conference. Berlin, Heidelberg: Springer; 2005. pp. 262-276

[23] Baader F, Horrocks I, Sattler U. Description logics as ontology languages for the semantic web. In: Mechanizing Mathematical Reasoning. Berlin, Heidelberg: Springer; 2005. pp. 228-248

[24] Richards T. Clausal Form Logic. An Introduction to the Logic of Computer Reasoning. Boston, MA, USA: Addison-Wesley Longman Publishing Co., Inc.; 1989

[25] Lukasová A, Žáček M, Vajgl M. Carstairs-McCarthy's morphological rules of english language in RDFCFL graphs. In: International Conference on Applied Physics, System Science and Computers (APSAC). Springer Lnee; 2016. pp. 169-174. DOI: 10.1007/978-3-319-53934-8_20

[26] Žáček M. Ferdiánová V. Solving logic problems with associative networks in the course of knowledge representation. In: International Conference on Industrial Technology and Management Science: Advances in Computer Science Research 2014 China. France: Atlantis Press; 2015. pp. 121-124. ISBN: 978-94-6252-123-0

Ontology Language XOL Used for Cross-Application Communication

Jinta Weng, Jing Qiu and Ying Gao

Abstract

The 2000s may be the flourishing time of the topic of ontology. Specialists and scholars concentrated to define ontology effectively and formulated uniform ontology protocol. Ontology language can be classified into SHOE, OML, XOL, OIL, OWL, and RDFs by different protocols and syntaxes. As for effective exchange of the different ontology messages in different applications, US bioinformatic community and researcher develop a XML-based ontology language. With the simplified OKBC-Lite protocol and flexible XML syntax, XOL offers the ways to define an ontology with the human-readable XML, simplified protocol, and compatible interface. In this chapter, we will introduce its motivation from history, orientation in development, semantic usage, and interpreted example in detail.

Keywords: ontology exchange language, Ontolingua, XOL, open knowledge base, Semantic Web

1. Introduction

Internet had maintained a rapid development between the 1990s and 2000s, which not only gives birth to various applications, abundant network facilities, and diverse websites but also accelerates the next generation of Semantic Web. After Berners-Lee put forward the imaginary structure of Semantic Web in 1998, W3C with many semantic work teams is dedicated to develop the technical standard of Resource Description Framework [1]. As ontology is the essence and basic of a resource, technical combinations of paradigm and languages are used to define it.

1.1 Background

Knowledge engineering has become an essential part of expert system in artificial intelligence. It is important to define the specific knowledge, also known as domain database or knowledge base, for multiple applications. However, traditional knowledge base just reveals the key and value of the data, thus paying less attention on ontology.

Ontology is the description and formulization of thing. By full-semantic and expressive ontology, more information and relationship are able to excavate. In order to build more humanistic and intelligent system, scholars had developed different ontology languages. Although many ontology languages give methods to solve the ontology definition. However, a new language or ontology protocol should also be formulated to deal with the cross-application problem.

1.2 Motivation

Accompanying with the development of Internet, more infrastructure, application, and knowledge base are generated. In normal knowledge supported systems, domain expert will first considerate the software environment and self-knowledge background and then choose the suitable knowledge scheme and ontology for the system. However, when it comes to the cross applications or large knowledge-assisted system, ontologies in system need to be reused. First, knowledge scheme in different systems may exist difference from the expert's personal cognize. Second, it offers several ontology languages for each system; thus, different ontology schemes can show in different formats, which make it hard to communicate in different applications. Third, the increasing demand of openness and the sharing lead of ontology could be exchange. Therefore, an ontology exchangeable protocol or new ontology language supporting to exchange should be redefined.

To realize the need of an evaluation on ontology in bioinformatics, several researchers on the US evaluation team developed a new specific ontology language—XOL [2]. By flexible XML expression and simplified protocol, XOL (xml-based ontology exchange language) is able to express and exchange different ontology information across incompatible applications.

1.3 Definition

XOL is an ontology language developing for exchange ontology in cross applications. It takes inspiration from OKBC (a protocol used for open knowledge base, see in Ref. [3]) and Ontolingua (another ontology used for reusing and editing ontology, see in Ref. [4]). Its syntax is based on human-readable and high compatible XML document. XOL can also respect as one effective intermedia language in ontologies' use, exchange, negotiation, and cocreation.

1.4 A simple example

Note the following XOL definitions:
```
<class>
<name> [class-name] </name>
</class>
<slot>
<[slot-attribute]></[slot-attribute]>
</slot>
<individual>
<name></name>
<type></type>
<...></...>
</individual>
```

All of above XOL elements are pertained to all ontologies. Between the pair of <class></class> defines the basic information of this ontology, like the name of the class during the tag pair of <name></name>.

Pair of <slot></slot> will depict the attribute and restriction of the class, like value's type of the attribute and the data restriction.

The last tag <individual> </individual> will give an instance of self-class or multiclass. It is not allowed to use the subclass as the individual element.

With the human-readable and self-defined XML syntax, XOL can express the ontology in a concise way. However, it may also lead to the ontology inconformity while using XML syntax merely or personally. A more restrictive and stationary tag OKBC-Lite was chosen soon.

2. Why is XOL based on XML?

Generally speaking, each ontology language makes up for using syntax and language protocol. To realize the essence of XOL, we will show the different classifications of ontology languages based on the syntax and semantic rules.

According to the use of syntax, we can classify the ontology languages into three types as follows:

HTML format: Hypertext Markup Language (HTML) is the basic document mark of the current web. To extend the semantic character of the HTML, ontology language like SHOE offers an effective way to support semantic annotation by more extended webpage label.

XML format: Extensive Markup Language (XML) is a more human-readable and concise document for storing and defining different data. Ontology document made by XML format can easily locate by its hierarchical structure and semantic DTD tag.

RDF format: Resource Description Framework (RDF) is a new way to define ontology after XOL. It is a resource model always accompanied by a specific URI and extended specific XML-like label to depict the relation and knowledge model between the resources. It not only specifically and strictly expresses the data but also makes the alternation, merging, and inference possible.

According to language protocol of these languages, ontology language can divide into first-order predicate logic language, frame-based language, and concept-role restriction language.

First-order predicate logic language is the most accurate and original language in knowledge representation. The predicate formula is the formula formed by joining some predicates together with the predicate join symbol, like the largest formalized language Cyclo [5] and KIF [6].

Frame-based language is a language that includes the aforehand defining framework and simplified first-order logic language. Owing to excessive strict first-order predict logic and unreadable syntax, Ontolingua and frame logic are developed to remedy this defect.

Concept-role restriction language is an effort that most language currently adopts. This type of language offers a hierarchy way to represent the hyponymy by concept and the individual's signal. It reveals the relationship and value restrictions between different ontologies by role mark, like OML [7].

To note the difference between ontology languages in multiple syntax formats, we will give a detailed introduction for some ontology language with the technical developing route.

2.1 SHOE (HTML format)

HTML had covered with a long history before the World Wide Web (WWW) appeared and is one of document standards of Standard Generalized Markup Language (SGML). SGML offers a high standard and complicated description about the document resource. As SGML is hard to learn, use, and realize, researcher put forward the HTML in 1989 after considering the computer's ability. HTML is the mere application of SGML in the WWW times. After few years of great development, HTML is widely known in the web document district. Semantic Web, as the next generation of WWW, is also the use of the HTML syntax in the ontology language. We called this simple HTML ontology language as SHOE.

SHOE is a specification that describes an extension to HTML, which provides a way to semantically describe important information about HTML or other web documents [1]. It offers a hierarchical classification mechanism for HTML documents and non-HTML documents or subsections of HTML documents. The intent of this specification is to make it possible for user agents, robots, and so on, to gather truly meaningful information about web pages and documents, enabling significantly better search mechanisms and knowledge gathering. Let us take the SHOE as an example; it can divide into two steps as follows:

Define an ontology

<Ontology "ontology-unique-name" version = "1.0" backward-compatible-with = "version list">

Use an existing XOL ontology

<Ontology-extends "ontology-unique-name" version = "Version" backward-compatible-with = "version list" URL = "location">

This is a simple way to define ontologies containing rules. Ontology simply means an ISA hierarchy of categories and a set of relations between these categories in this SHOE specification. Categories will also inherit relations defined in parent categories. However, this specification does not as yet define any other forms of relationships (transitive closures, inverses, negations, etc.) and use the complicated and human-unreadable Hypertext Markup Language as the basic syntax.

2.2 KIF

At the same time, first knowledge interchange format was proposed by American National Standard (dpANS). Though many ontology languages are still developed, researcher in Standard University begins to design a language for the intercommunication. Interchange of knowledge or ontology thought out disparate computer systems (different programmers, different languages, and other discrepancy in interknowledge sharing). KIF language is logically comprehensive with declarative semantics.

In addition to these essential features, KIF is designed to maximize the implementability and readability. KIF provides a declarative language for describing knowledge. As a pure specification language, KIF does not include commands for knowledge base query or manipulation.

2.3 GFP

GFP is first motivated by the hierarchic framework design of frame-based knowledge representation systems (FRSs) used at the Stanford Knowledge Systems Laboratory for accessing Cyc, KEE, and Epikit [8]. FRSs can contain all of the database systems and knowledge systems or other frame-like projects. It is complementally developed to support knowledge sharing. It specifies a new protocol, Generic Frame Protocol (GFP), for connecting knowledge bases (KBs) in FRSs. In more detail, it provides numbers of operations to formulate a general interface for all of the FRSs. Also, complementary tools were also produced to keep independent and general operation generation. GFP shows well compatibility between different languages, including Java, C (client implementation only), and Common Lisp. Thus, some format conversations of languages are also needed.

2.4 OKBC

After GFP coming out, OKBC, a new protocol called Open Knowledge Base Connectivity, has taken it up in more implicit knowledge model and knowledge operation. It first uses some open ontology systems, such as EcoCyc, GKB Editor, and Ontolingua projects. With 2 years of development, OKBC quickly used in several ontology sharing projects.

OKBC handles the knowledge in more implicit representation formalism, which we called OKBC Knowledge Model in later years. This model not only supports an object-oriented representation of knowledge but also can found the commonly knowledge structure from different KRSs. Therefore, it can serve as an implicit knowledge interlingua by its powerful character in knowledge for all of the systems using OKBC.

2.5 Ontolingua

Ontolingua, which accompanies with different ontology languages breaking out, can serve as a basic framework to support open or domain knowledge sharing system. The syntax of Ontolingua definition is based on GFP. It is motivated by the need of Summer Ontology Project, a pilot study in which researchers from several groups and institutions met weekly to design ontology of terms used in modeling electromechanical devices. **Figure 1** depicts the structure of Ontolingua.

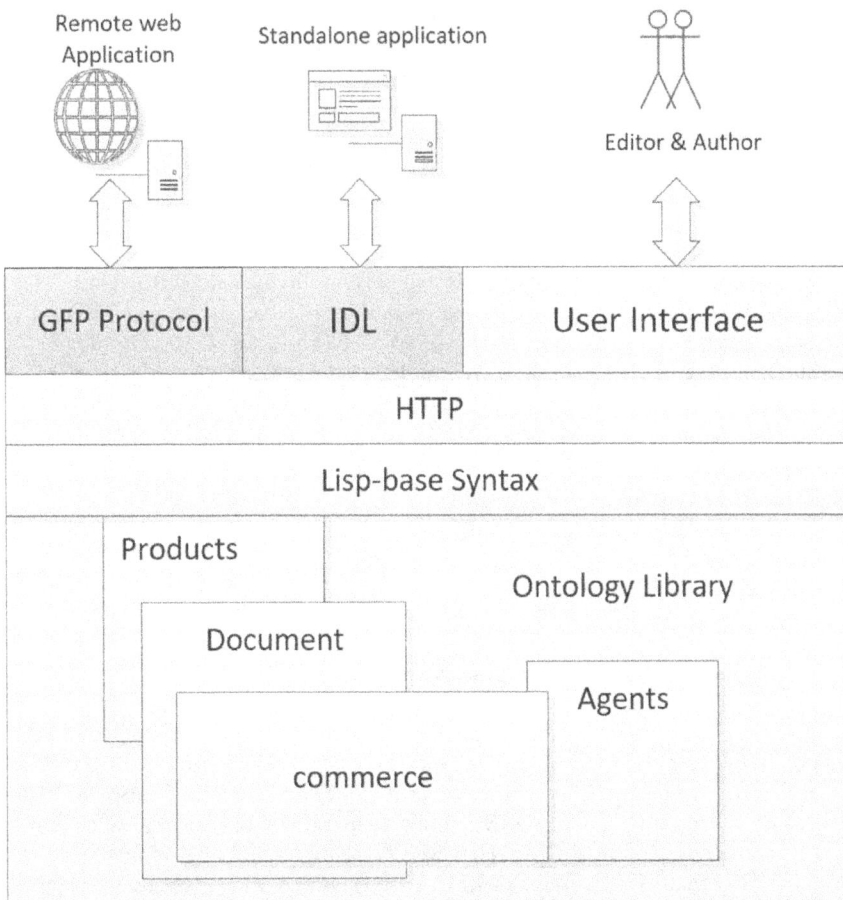

Figure 1.
The structure of Ontolingua.

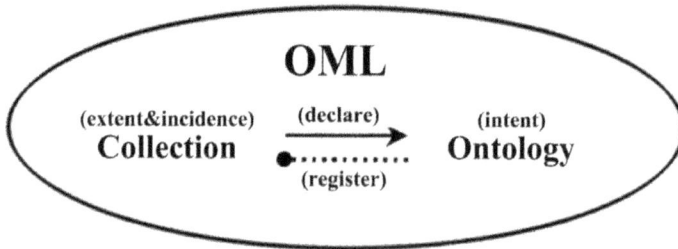

Figure 2.
The structure of OML.

At first, authors and editors will publish their ontology or maintain an ontology with a HTTP protocol connection. Users may access the different defined ontologies in Ontology Library, for example, ontologies, product ontologies, document ontologies, commerce ontologies, and agent knowledge. Beside the ontology editors or editors, some remote application may also need to duly connect this ontology library in GFP protocol to exchange or manipulate ontology. When it comes to standalone application, interface definition language with a specific and functional method will retrieve an ontology batch from ontology library. Ontolingua keeps a balance in generalization expressed GRP protocol and selectivity in some standalone application by own interface definition language (IDL).

Ontolingua can thereby be shared by multiple user and research groups using their own favorite representation systems and can be easily ported from system to system. The syntax of Ontolingua definition is simplified with some class name, argument, and documentation string.

2.6 OML

XOL is more similar to Ontolingua. However, Ontolingua using OKBC is frame-based design and less semantic expression. At the same time, a separate set of researchers is pursuing a concept-role restriction language—OML. Ontology markup language (OML) is an adaptive change language based on SHOE. Earlier versions of OML were basically an Extensive Markup Language (XML) translation of the SHOE language with suitable changes and improvements. Common elements existing in OML can be described by paraphrasing the SHOE documentation to some degree. Now OML is highly RDF Schemas compatible, although it has own solution within the namespace problem. More importantly, OML has incorporated own elements and expressiveness of conceptual graphs. As shown in **Figure 2**, by declaring and registering operation, OML can be seen as a bridging connection between Ontology and Collection, which reflects more extent and incidence of ontology.

2.7 An ontology exchange language in XML format: XOL

With the development of the relative language, protocol and definition, and the need of an Evaluation of Ontology Exchange Languages for bioinformatics, several researchers on the evaluation team are currently developing a specification of XML expression of Ontolingua using OKBC, while a separate set of researchers is pursuing a frame-based version of OML. However, Ontolingua first uses a Lisp-based syntax (rather than HTML-based or XML-based), which leads to become hard to develop and maintain, though the semantics of OKBC-Lite are extremely similar to the semantics of Ontolingua. At this background, XOL was first published in 1999.

3. The usages of XOL

The usages of XOL are based on frame-based approach. In this part, we will introduce the constituent part and its usage mode.

3.1 Basic data type

- Integer

- Floating point numbers

- Double-precision floating point numbers

- Strings

- Boolean

- Name of class

3.2 Classes

Classes are composed of entities. Entity that is not the class but an instance of a class or multiple classes is said to be Individual. Classes and Individual distinguish by whether entity is a class of another entity or not. Class entity male or class entity female is the subclass of another class entity human, a man called Joe is an Individual entity of the class entity. The following Class is the basic Class description defined in the OKBC-Lite (**Table 1**).

3.3 Slots on slots

Slot is common property of each class or instance. The attribute "Documentation" of class has an introduction to this class. When it comes to the specific KB, slot may divide into "own slot" in this KB or "template slot" inheriting from class.

Slots on slots, as shown in **Table 2**, are the several restrictions or declarations defined to this slot. Although it may be inherited from other KB or class, restriction or declaration on slots in this XOL file is exclusive.

Name	Description	Name	Description
THING	The root of the class hierarchy The superclass of every class	SYMBOL	The class of all symbols A subclass of THING
CLASS	The class of all classes	INDIVIDUAL	The class of all entities that are not classes
NUMBER	The class of all numbers A subclass of INDIVIDUAL	INTEGER	A subclass of NUMBER
STRING	The class of all text strings A subclass of INDIVIDUAL	LIST	The class of all lists A subclass of INDIVIDUAL

Table 1.
The classes of XOL [2].

Name	Function
DOMAIN	Specifies the domain of the binary relation represented by a slot frame
SLOT-VALUE-TYPE	Specifies the classes of which values of a slot must be an instance
SLOT-INVERSE	Specifies the inverse relation for a slot
SLOT-CARDINALITY	Specifies the exact number of values that may be asserted for a slot for entities in the slot's domain
SLOT-MAXIMUM-CARDINALITY	Specifies the maximum number of values that may be asserted for a slot for entities in the slot's domain
SLOT-MINIMUM-CARDINALITYNUMERIC	Specifies the minimum number of values for a slot for entities in the slot's domain
SLOT-NUMERIC-MAXIMUM	Specifies a lower bound on the values of a slot for entities in the slot's domain
SLOT-NUMERIC-MINIMUM	Specifies an upper bound on the values of a slot for entities in the slot's domain
SLOT-COLLECTION-TYPE	Specifies whether multiple values of a slot are to be treated as a set, list, or bag

Table 2.
The slots on slots of XOL.

Name	Description	Name	Description
VALUE-TYPE	Value can be class or multiclass or set of value	MINIMUM-CARDINALITY	The class of all symbols A subclass of THING
INVERSE	Describe the slot relation is reverse and value is reverse slot	NUMERIC-MINIMUM	Specifies lower bound on the number-type values of slot
CARDINALITY	Specifies the exact number of values asserted for a slot	NUMERIC-MAXIMUM	Specifies upper bound on the number-type values of a slot
MAXIMUM-CARDINALITY	Specifies the maximum number of values asserted for a slot	COLLECTION-TYPE	Specifies whether multiple values of a slot are to be treated as set/list/bag.

Table 3.
The facet of XOL.

3.4 Acceptable facet

Different with slots on slots, facet is a restriction on the value of slot of individual. For instance, facet VALUE-TYPE and facet NUMERIC-MINIMUM describe the type of value of a slot and the minimum of value of a slot. **Table 3** is the acceptable facet defined in OKBC-Lite.

Facet can also divide into two parts:

- Own Facet (only state on current class or current KB)

- Template Facet (inherit from another class)

4. XOL example

Every XOL file must start with the following XML tab in the beginning.
<? xml version="1.0"?><!DOCTYPE module SYSTEM "module.dtd">
As the whole KB's first description, 'module section' will illuminate some information (name, type of DB, in which package, etc.) about this KB.
<Module> /*Every XOL file will start with a module mark */
<Name>name of this KB</Name>
<kb-type>which existing kb type</kb-type>
<package>self-defined package name</package>
The second section is 'class section.' In this section, we will introduce all these KB's classes or inherit from other class by the tag 'subclass-of' or 'instance-of.'
<Class>
<Name>name of class</Name>
<Documentation>String-type description</Documentation>
[Other-option-slot] (Subclass-of | instance-of | etc.)
</Class>
The third section is 'slot section,' which declares all slots in the class and slots existing slot value in an individual, such as the class-name, class-documentation, and other-option slot.
<Slot>.
<Name>name of own slot or template slot</name>
<Documentation>description</documentation>
<Domain (or other slots on slots in **Table 2**)>.</Domain>
</Slot>
The last section generally is 'individual section.' It contains all instances and their values in each slot. Also, it declares restriction of value of slot and slot-values.
<Individual>
<Name></name>
<instance-of></instance-of>
<Slot-values>
<Name>... </name>
<Values>...</values>
<value-type (or facet in **Table 3**)>...</value-type>
</slot-values>
</individual>
At last, remember that XOL file must use </model> to note the ending.

5. Future developments: OIL

Framework representation is to express the concepts, instances, classes, and relationships used in ontology in the form of framework. XOL is such a framework method-based ontology representation language. Unlike the rich expressions in logic-based approach, XOL leads to the deficiency in reasoning ability. The main differences stem from the fact that frames generally provide quite a rich set of primitives but impose very restrictive syntactic constraints on how primitives can be combined and on how they can be used to define a class.

Due to the deficiency of XOL in grammatical reasoning and the continuous development of DL notation, another new ontology interactive language OIL is defined [9]. It is not only an ontology description language but also a frame-based web language and an XML and RDF compatible ontology language. Its appearance unifies the characteristics of traditional ontology language and endows the new object into the inference layer.

6. Conclusion

The emergence of XOL was inspired by existing ontology and protocol, for example, SHOE, KIF, GFP, OKBC, and so on. XOL is a bridge language, which let the ontology using frame-based approach can be expressed in a simplifier way during the XML file. By the human-readable XML and unified Label limitation, it can use as an ontology exchange language during the cross application, which allows to obey the use of the XOL (in fact is OKBC-Lite) restriction. However, lacking of inferential capability and more logical expression, XOL was replaced by the subsequent ontology language OIL considering the multilanguage and more logic restriction that enable to validate ontology.

This chapter gives an overlook of XOL from the historical development across the different ontology languages. Note that XOL is not the first language in defining the ontology language for exchange date. It merely complements XML syntax and uses a simple frame-based OKBC protocol. However, it still lacks more compatible with multiple ontology protocols and different syntax. Without the more consideration into inference, ontology quickly would replace by stronger ontology system.

We also found that while designing a better widely used ontology language, we should keep a right balance between generality and specificity or between compatibility and limitation. We will focus more on the result comparison between different ontology methods and the humanity background within different languages.

Author details

Jinta Weng, Jing Qiu* and Ying Gao
Guangzhou University, China

*Address all correspondence to: qiujing.ch@gmail.com

IntechOpen

References

[1] Hendler J, Berners-Lee T. From the semantic web to social machines: A research challenge for AI on the world wide web. Artificial Intelligence. 2010;**174**(2):156-161. DOI: 10.1016/j. artint.2009.11.010

[2] Karp R, Chaudhri V, Thomere J. XOL: An XML-based ontology exchange language. Version 0.4. 2002. Available from: http://www.ai.sri.com/~pkarp/xol [Accessed: 18 May 2019]

[3] Chaudhri VK, Farquhar A, Fikes R, Karp PD, Rice JP. OKBC: A programmatic foundation for knowledge base interoperability. In: Proceedings of the National Conference on Artificial Intelligence. 1998

[4] Gruber TR. Ontolingua: A mechanism to support portable ontologies. Knowledge Systems Laboratory. Stanford, CA: Computer Science Department, Stanford University; 1992:94305

[5] Lenat DB, Guha RV. Building Large Knowledge-Based Systems; Representation and Inference in the Cyc Project. 1st ed. USA: Addison-Wesley Longman Publishing Co.; 1989. DOI: 10.5555/575523

[6] Michael R. Genesereth. Knowledge interchange format [Internet]. 1998. Available from: http://logic.stanford.edu/ kif/dpans.html [Accessed: 18 May 2019]

[7] Kent RE. Conceptual knowledge markup language: An introduction. NETNOMICS. 2000;**2**(2):139-169. DOI: 10.1023/a:1019186729572

[8] Vinay K. Chaudhri, et al. Generic Frame Protocol (GFP). 1997. Available from: http://www.ai.sri.com/~gfp/ [Accessed: 18 May 2019]

[9] van Harmelen F, Horrocks I. FAQs on OIL: The ontology inference layer. IEEE Intelligent Systems. 2000;**15**(6):69-72

Chapter 4

An Ontology-Based Approach to Diagnosis and Classification for an Expert System in Health and Food

Friska Natalia, Dea Cheria and Santi Surya

Abstract

In this chapter, we will discuss how to make an ontology-based expert system easy to use and apply to community sustainability issues without pay. Ontology itself plays an essential role in the diversity of knowledge and management methods that can simplify communication between expert domains and users. The scope of this study is health and food, which is expected to help people realize the disturbances they experience. In this chapter, we will discuss two cases: (i) determine the depressive disorder a person has based on their health condition and (ii) determine the type and variant of rice according to needs. Ontology is a method used in research that can be structured and systematic real-world representation that is equal and provides a reference model. The results of this study are an expert system model and mobile applications to help users overcome the problems in the health and food fields with the ontology method. The objective of this study is to develop the application based on the ontology method to make it easy for people to find information on expert systems.

Keywords: expert system, modelling, ontology, depressive disorder, variant of rice

1. Introduction

Ontology matching is a field of research that is in high demand today, where information exchange and reuse of knowledge are essential topics in the development of the ontology; this is one solution to the problem of semantic heterogeneity. Matching ontology aims to find correspondence between entities in the semantic ontology. In this study, we will discuss how to make an ontology-based expert system to be easily used by the community without pay. Ontology itself plays a vital role in the diversity of knowledge and the way it is regulated. Ontology is a structured and systematic equivalent real-world representation. Moreover, ontology also provides a referral model that can simplify communication between expert domains and improve understanding and information sharing. There are several previous research related to ontology. The first paper describes a knowledge system for improving RFID recognition by using fuzzy ontology [1]. The second paper shows the fuzzy ontology with fuzzy concepts is an extension of the domain ontology with crisp concepts [2]. The third paper concerns to integrate ontologies from food, health, and nutrition domains to help the personalized information systems to retrieve food and health recommendations based on the user's health conditions and food preferences [3]. The fourth paper provides advancement to the research of diabetes diagnosis using CBR algorithm [4]. The fifth paper proposes a search based on multiple ontologies to make information retrieval efficient. It rewrites the user query by adding semantic information, after consulting multiple ontologies [5]. The other paper is to integrate ontologies from food, health, and nutrition domains to help the personalized information systems to retrieve food and health recommendations based on the user's fitness conditions and food preferences [6].

In this study, the object of research to be used is a ubiquitous object in the community, namely, health and food. Ontology itself plays an essential role in the diversity of science and how to regulate it. Cytology is a real structured and systematic representation of the world. Moreover, ontology also provides a reference model that can simplify communication between domain experts and improve understanding and information sharing.

Many similar studies have used the decision tree so that its use is more familiar than ontology. However, in this study, classification ontology does not require elimination or precise calculations to be able to take a conclusion, as did the decision tree, because the results are taken from predetermined criteria. Although ontology cannot be flexible in choosing criteria for different outcomes, the classification ontology method is more directed at relationships within each entity rather than elimination based decision-making.

(i) Health is a very complicated problem, which is the result of various natural and human-made environmental problems. The coming of a disease is something that cannot be rejected, although sometimes it can be prevented or rejected. The concept of health or sickness is truly undivided and universal because other factors influence greatly, especially sociocultural factors. Therefore, it is crucial to have thoughts about the concepts of health and illness. If the idea of the concept is based on the correct concept, then the community will also find the right alternative to solve their health problems. The type of disease in this world is comprehensive. Some include common diseases, but some include diseases that are quite difficult to avoid. One of them is an allergic disease. (ii) Food will be discussed for the selection of rice to be consumed in fulfilling daily intake depending on consumer tastes. Rice is one of the essential cereals needed by humans for consumption. Even though some countries in the world have basic needs to switch to wheat products, the Indonesian people still rely heavily on rice as a basic necessity to meet their daily needs.

The results of this study are an expert system model and application in the form of a mobile application using ontology created by Protégé ontology web language (OWL) to help users. This research is also expected to help users know the types of depressive disorders and recognize the criteria of someone experiencing the depressive disorder in the health sector and help to sort and select the variants and types of rice that best suit the needs of users with the ontology method. The method that will be used is ontology because ontology provides potential compatibility and chooses the results that are most suitable according to the needs and criteria given. Also, ontology is a way to describe the meaning and relationship of a term. The description can help computer systems to understand the terms used more efficiently. Thus the needs can be sorted and chosen appropriately without just perception.

Based on the above problems, these are the objectives of this study:

1. Know the application of the ontology method in making expert systems.

2. Make it easy for people to find information on expert systems.

2. Problem statement and research methodology

In this study, there are two scenarios to be discussed: health and food.

i. Health: The Indonesian people still lie about mental disorder, so what usually happens is that they isolate patients. Activities, social life, the rhythm of work, and relationships with families are disrupted due to symptoms of anxiety, depression, and psychosis. Someone with any mental disorder must get treatment immediately. Delay in treatment will be more detrimental to sufferers, families, and society. In this study, mental health ontology (for cases of depressive disorder) will be made based on the types and provide any information that characterizes a person suffering from a depressive disorder. The design will use Protégé, with the latest data made by the American Psychiatric Association and implemented into an Android-based mobile application, which is expected to help people realize the disturbances they experience and get immediate treatment.

ii. Food: In this scenario in the selection of rice, there is no specific reference for making choices. Difficulties are also felt when there are no specific benchmarks based on experts or the right knowledge to determine the compatibility between processed and rice types. Besides, the application of ontology can also be applied in the world of food and medicine. This makes it interesting to apply the ontology to one aspect of food, in this study, rice, intending to create an Android-based model and expert system application to sort and select the variants and types of rice that best suit the needs of users with the ontology method. Because ontology provides potential compatibility and choices, the results that are most suitable according to the needs and criteria are given. Thus the needs can be sorted and chosen appropriately, not only with perception.

The method used in this study is the classification ontology method; besides because it is the method most often used in similar studies, this method shows more the real identity and relationship in each entity compared to the decision tree which leads to eliminating unnecessary calculations. Ontology consists of basic terms, the relationship between those terms, and rules that incorporate them. The ontology can become knowledge that can be shared and used in multiple applications. The reason for using this method is because it is the most suitable way to perform data groupings and interclass entity relationships. Considering the object of this study, then the method used in this study is the classification ontology method because this study does not require the elimination of particular calculations and the classification ontology method is more directed at relationships in each entity rather than elimination-based decision-making.

In this research, the rapid application development (RAD) system development method will be used to develop mobile applications because this method is commonly used for making relatively short systems. The stages in the rapid application development method are [7–9]:

1. Stage requirement planning

   At this stage, a plan is carried out to determine what data is needed for system requirements.

2. Stage design

   At this stage, a temporary design is made that focuses on presentation.

3. Implementation phase

At this stage, the results will be translated into the appropriate programming language and then be tested before being disseminated.

3. Design and implementation

3.1 First scenario

Requirement planning, in the system requirement: the system can select for user data search purposes, and the system can do searching or filtering to find the right data. In data requirement: All data were from the *Diagnostic and Statistical Manual of Mental Disorders Fifth Edition*, which was created by a selection of selected psychiatrists from the American Psychiatric Association. In this study, only a limited part of the depressive disorder and diagnostic criteria in it.

Depressive disorder is the presence of sad, empty, or irritable mood, accompanied by physical and cognitive changes that impact function [10]. Based on that, we can categorize it into eight types: Someone who has disruptive mood dysregulation disorder does not experience depressed mood, but becomes more irritable and more sensitive, often has problems with his mood, has experienced symptoms for more than 12 months, is not affected by drugs, has never had medical treatment, is not coming month (for women), and does not have other psychological diagnoses. Also, the person exhibits persistent irritability and frequent episodes of extreme verbal and behavioral dyscontrol toward people or property out of proportion to the situation and is inconsistent with developmental level occurring on average three or more times per week. Someone who has major depressive disorder experiences depressed mood, does not become more irritable and more sensitive, often has problems with his mood, has experienced symptoms for more than 2 weeks, is not under the influence of drugs, has never had medical treatment, is not coming month (for women), and does not have other psychological diagnoses. Someone who has persistent depressive disorder experiences depressed mood, does not become more irritable or more sensitive, often has problems with his mood, has experienced symptoms for more than 2 years, is not under the influence of drugs, has never had medical treatment, is not coming month (for women), and does not have other psychological diagnoses. Someone who has premenstrual dysphoric disorder experiences depressed mood, becomes more irritable or more sensitive, often has problems with his mood, does not know when to start feeling the problem, is not under the influence of drugs, has never had medical treatment, is on the moon (for women), and does not have other psychological diagnoses. In all of the cycles, symptoms present in the final week before menses, start to improve within a few days after onset, and become minimal or absent in the week postmenses. Someone who has substance-/medication-induced depressive disorder experiences depressed mood, does not become more irritable or more sensitive, often has problems in his mood, has experienced symptoms for more than 1 month, is in the influence of drugs, is undergoing medical treatment, does not have moderate menstruation (for women), and does not have other psychological diagnoses. Someone who has depressive disorder due to another medical condition experiences depressed mood, does not become more irritable and more sensitive, often has problems with his mood, does not know when to start feeling the problem, is not influenced by drugs, has undergone medical treatment, is not having menstruation (for women), and does not have other psychological diagnoses. Someone who has other specified disorder and unspecified depressive disorder experiences depressed mood, becomes more irritable or more sensitive, rarely has problems with his mood, has experienced symptoms within a period of time but not daily, is not influenced by drugs, has never had medical treatment, is not having a period (for women), and maybe having another psychological diagnosis.

Table 1 shows the choice of whether the user experiences a depressed mood. Depressed mood includes poor appetite or overeating, insomnia or hypersomnia, low energy or fatigue, low self-esteem, difficulty in concentrating or difficulty in making decisions, despair, and anxiety.

In Table 2, the more the user is accessible to anger means the user is more accessible to emotion than he should. It is easier to get angry because he is more sensitive to ordinary things.

Table 3 shows the problem in the mood is divided into three, such as, all the time which means the problem in the mood is experienced at any time; often which means the problem in the mood is experienced almost every day, but not every time; while rarely can mean the problem in the mood is only experienced occasionally and not necessarily every day, or only at certain times.

Table 4 describes the effect of narcotics on users. If a user is a drug user when feeling symptoms of depressive disorder, then the user may be included in substance-/medication-induced depressive disorder or even not included in any depressive disorder.

Medical treatment can be seen in Table 5. If the user is undergoing treatment, there is a possibility that the user will enter into substance-/medication-induced depressive disorder, or if the user has undergone treatment, there is a possibility

Number	Disorder name	Depressed mood
1	Disruptive mood dysregulation disorder	No
2	Major depressive disorder	Yes
3	Persistent depressive disorder (dysthymia)	Yes
4	Premenstrual dysphoric disorder	Yes
5	Substance-/medication-induced depressive disorder	Yes
6	Depressive disorder due to another medical condition	Yes
7	Other specified depressive disorder	Yes
8	Unspecified depressive disorder	Yes

Table 1.
Depressed mood.

Number	Disorder name	Easy to get angry
1	Disruptive mood dysregulation disorder	Yes
2	Major depressive disorder	No
3	Persistent depressive disorder (dysthymia)	No
4	Premenstrual dysphoric disorder	Yes
5	Substance-/medication-induced depressive disorder	No
6	Depressive disorder due to another medical condition	No
7	Other specified depressive disorder	Yes
8	Unspecified depressive disorder	Yes

Table 2.
Easy to get angry.

Number	Disorder name	Mood problem
1	Disruptive mood dysregulation disorder	Often
2	Major depressive disorder	All the time
3	Persistent depressive disorder (dysthymia)	All the time
4	Premenstrual dysphoric disorder	Often
5	Substance−/medication-induced depressive disorder	Often
6	Depressive disorder due to another medical condition	Often
7	Other specified depressive disorder	Rarely
8	Unspecified depressive disorder	Rarely

Table 3.
Mood problem.

Number	Disorder name	In the influence of drugs
1	Disruptive mood dysregulation disorder	No
2	Major depressive disorder	No
3	Persistent depressive disorder (dysthymia)	No
4	Premenstrual dysphoric disorder	No
5	Substance−/medication-induced depressive disorder	Yes
6	Depressive disorder due to another medical condition	No
7	Other specified depressive disorder	No
8	Unspecified depressive disorder	No

Table 4.
The influence of drugs.

Number	Disorder name	In medical treatment
1	Disruptive mood dysregulation disorder	No
2	Major depressive disorder	No
3	Persistent depressive disorder (dysthymia)	No
4	Premenstrual dysphoric disorder	No
5	Substance−/medication-induced depressive disorder	Is undergoing
6	Depressive disorder due to another medical condition	Ever undergoing
7	Other specified depressive disorder	No
8	Unspecified depressive disorder	No

Table 5.
Medical treatment.

that the user will enter into depressive disorder due to another medical condition, or even depressive disorder.

Table 6 is only for women, and if it meets the criteria, there is a possibility that the user includes premenstrual dysphoric disorder, but it may not be included in any depressive disorder.

Table 7 has shown the length of time for how long a person experiences symptoms of depressive disorder, starting from 2 weeks, 1 month, 12 months, 2 years, not long ago for those who experience infrequently, or do not know, because they feel they have enough old but only certain moments.

Table 8 describes that one of the main requirements in depressive disorder is not having another psychological diagnosis. Several other mental disorders have characteristics similar to depressive disorder; if the user feels that he has another psychological diagnosis, then maybe the user is included in other specified

Number	Disorder name	In the menstruation time
1	Disruptive mood dysregulation disorder	No
2	Major depressive disorder	No
3	Persistent depressive disorder (dysthymia)	No
4	Premenstrual dysphoric disorder	Yes
5	Substance–/medication-induced depressive disorder	No
6	Depressive disorder due to another medical condition	No
7	Other specified depressive disorder	No
8	Unspecified depressive disorder	No

Table 6.
In the menstruation time.

Number	Disorder name	How long
1	Disruptive mood dysregulation disorder	12 months
2	Major depressive disorder	2 weeks
3	Persistent depressive disorder (dysthymia)	2 years
4	Premenstrual dysphoric disorder	Do not know
5	Substance–/medication-induced depressive disorder	1 month
6	Depressive disorder due to another medical condition	Do not know
7	Other specified depressive disorder	Not long ago
8	Unspecified depressive disorder	Not long ago

Table 7.
How long.

Number	Disorder name	Have other diagnoses
1	Disruptive mood dysregulation disorder	No
2	Major depressive disorder	No
3	Persistent depressive disorder (dysthymia)	No
4	Premenstrual dysphoric disorder	No
5	Substance–/medication-induced depressive disorder	No
6	Depressive disorder due to another medical condition	No
7	Other specified depressive disorder	Could have
8	Unspecified depressive disorder	Could have

Table 8.
Have other psychological diagnosis.

Number	Rice type	Description
1	White rice	White rice is commonly consumed
2	Dark Brown Rice	Similar to white rice but slightly brownish
3	Brown rice	Has a reddish outer layer
4	Black rice	Rice that is rather blackish
5	White glutinous rice	Glutinous rice that is thick and white
6	Black glutinous rice	Blackish sticky rice
7	Parboiled rice	Rice that is soaked in warm water first
8	Mixed rice	A mixture of several types of rice
9	Basmati rice	Middle Eastern rice
10	Instant rice	Rice that quickly turns into rice
11	Japanese rice	Rice for Japanese cuisine (more springy)

Table 9.
Brief description of rice types.

depressive disorder or unspecified depressive disorder or even not including depressive disorder.

3.2 Second scenario

In order to create a system, a thorough analysis of system requirements is needed to make a great system. The analysis was carried out on interview data that had been conducted so far to produce a proper application which ran smoothly according to the initial needs. In this scenario, interviews were conducted to obtain the data needed. From the results of interviews conducted with the three speakers, one of the resource persons summarized the knowledge of the three experts, including himself, to make a rule in determining the type of rice that matches the processed food, along with the brand according to the specified criteria. Criteria are determined based on information from experts regarding the type of rice according to processing. **Table 9** shows the types of rice found in Indonesia based on the interviews.

The following are brands of rice that sell these types:

a. Myrice

b. Parakeet

c. Basket

d. Louhan

e. Goldrice Red

f. Goldrice Green

g. King rice

h. Swallow

i. VIP

j. Panda

k. Three guava

l. Cap

m. Penguin

n. BMW

o. Guci Mas

p. Flower Stamp

The following are processed rice commonly made by consumers in the rice stores according to the seller's knowledge:

a. Bakcang

b. Porridge

c. Black glutinous porridge

d. Gyudon

e. Egg-crust

f. Lemper

g. Lontong/ketupat

h. Rice

i. Baked rice

j. Fried rice

k. Gudeg rice

l. Corn rice

m. Yellow rice

n. Liwet rice

o. Team rice

p. Uduk rice

q. Sushi

Table 10 shows the criteria for rice used:

The following is a table of matches between the types, criteria, and brands of rice with processed rice; the original table sent by the guest speaker is in the Appendix.

Explanation of **Table 11**:

G = Glutinousness. (+) means it is more sticky or contains water. (−) means it is slightly watery or sticky.

L = Length. (+) means more oblong. (=) means more rounded.

A = Aroma. (+) means more fragrant. (−) means less fragrant.

R = Taste. (+) means sweeter. (−) means more acidic.

= means having these criteria within normal limits.

Number	Criteria	Description
1	Glutinousness	Rice stickiness level
2	Taste	The resulting taste (more acidic or sweet)
3	Aroma	The fragrance level of rice
4	Length	The length and shape of rice (somewhat oval or rather round)

Table 10.
Brief description of rice criteria.

Number	Processed	Types of rice	G	L	A	R	Brand of rice
1							
2	Bakcang	White rice, glutinous rice	+			√	Myrice, Penguin
3	Porridge	White rice			+	√	Flower Stamp
4	Black sticky rice porridge	Black glutinous rice				√	Parakeet, Bakul
5	Gyudon	Mixed rice, Japan	+			√	Louhan, Goldrice Green
6	Egg-crust		−	−			Swallow
7	Lemper	White glutinous rice	+		−		Swallow, King rice, Goldrice Green
8	Lontong/ketupat	White rice	+			√	Myrice, King rice
9	Rice	(All)	√	√	√	√	(All)
10	Roasted rice	White rice, parboiled rice, brown rice			+	√	Goldrice Red, VIP
11	Fried rice	White rice, red rice	−			√	Panda, BMW, Penguin
12	Gudeg rice	White rice	+			√	Three Guava, Capit
13	Corn rice	Basmati rice	+		+	√	Three Guava, Capit
14	Yellow rice	White rice	−		−	√	Panda, BMW, Penguin
15	Liwet rice	White rice	+			√	Three Guava, Capit
16	Team rice	White rice	+			√	Three Guava, Capit
17	Uduk rice	White rice	−			√	Panda, BMW, Penguin
18	Sushi	Japanese rice, mixed rice	+			√	Louhan, Goldrice Green

Table 11.
Processed compliance tables with types, criteria, and brand of rice.

4. Results

Figure 1 describes a flowchart design used in the process of making this system or application. First, the collected data will be analyzed and will be used as a reference in making classes and subclasses on ontology, and the process of creating classes and subclasses will involve the use of Protégé tools. Then the next step is to determine the property. At this stage, we will determine the property object and data property, which will be needed as attributes and relations of each data. The first property object is created based on the class and subclass that were created before, and each class will have its data property.

After that create a data property; this time the creation will be affected by property objects, and this data property will be useful to name the class and the data to be included in this ontology because each data will have its name used for identifying it. Then classify all data entered into ontology. Each incoming data must have at least one relationship with other data so that it can be used based on the relationship they have. All data will be given a relationship with each other; after that the data is ready and stored in the form of an OWL file, which will be used later in the application. Next is creating a SPARQL query that will be used to retrieve data from the OWL file. To be able to make the query, PREFIX must first be specified, which is the name of the place of the data [11]. Furthermore, the WHERE is determined, to give a limit on the data to be taken, by determining the conditions or conditions that must be met to retrieve the data. Inside the WHERE, there is a FILTER, which is useful for classifying data retrieval as needed. The next step is to do the making of an application, starting from the design of how the application will look up to the functions in the application. Besides that, it also makes a connection between the OWL file containing the ontology data and the application.

Ontology graphics, or commonly referred to as OntoGraf, is one of the features found in Protégé tools. This feature was introduced to the user, starting from version 4.3 [12]. The Protégé software adapts the Java programming language which can be customized according to user needs [13]. Usually, ontology research

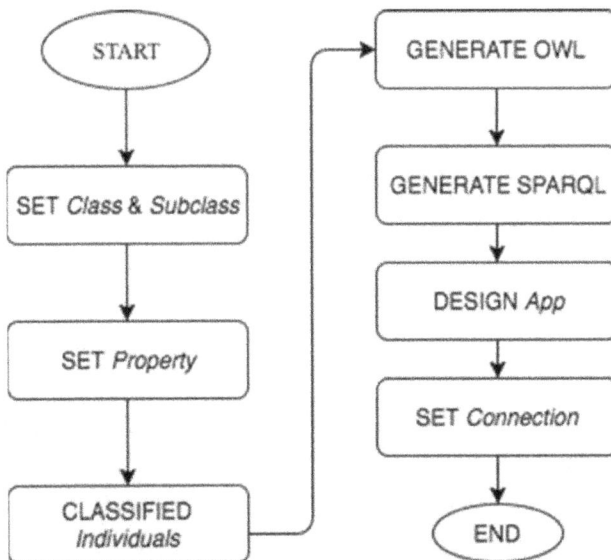

Figure 1.
Flowchart.

and acquired knowledge using Protégé software. The function of OntoGraf itself is to provide an interface so that users can manage navigation from relationships between classes, properties, and individuals contained in OWL files [14]. By using this menu, the user can see the display of relations between class, property, and individuals in the OWL file. There is an OntoGraf feature that provides interactive support for navigating the OWL ontology relationship. Various layouts are supported to regulate the structure of the ontology [15]. The types of data relationships supported include subclass, individual, domain or range object properties, and equivalence.

Figure 2 is an ontograph wherein the Protégé application; there are seven depressive disorders and eight subclasses of depressive disorder. Other specified depressive disorder and unspecified disorder is being one part because they have the same characteristics. The picture shows the depressed mood subclass has two individuals, namely, not experiencing and experiencing. The subclass of psychological diagnosis has two individuals, namely, none and possible; subclass how long describes how long the user has experienced problems such as depressed mood or more irritable, that is, more than 12 months, more than 2 weeks, do not know, more than 1 month, not long ago, and more than 2 years.

Figure 3 is the first-level display of the ontograph. The conclusion is that rice has members, namely, the type of rice that exists. Then as a subclass of rice, there are descriptions and dishes. The description has a brand and variable, while the dishes contain rice dishes. Next level 2 of the ontograph is a member of each subclass. Rice dishes consist of names of processed foods, and brands have members, namely, existing rice brands, while variables have members, namely, the criteria for rice used.

The OWL file is the output generated from the design that has been done before, namely, the design of classes, subclasses, property objects, and datatype properties. The following are the results of the design. The class in the class design is depressive order. **Figure 4** shows that the depressive order class has subclasses, which are how long, depressed food, psychological diagnosis, food problems, problems with tempers, menstruation, drugs, and medicinal medicine, while **Figure 5** shows the contents of the class and subclass of OWL in the second scenario.

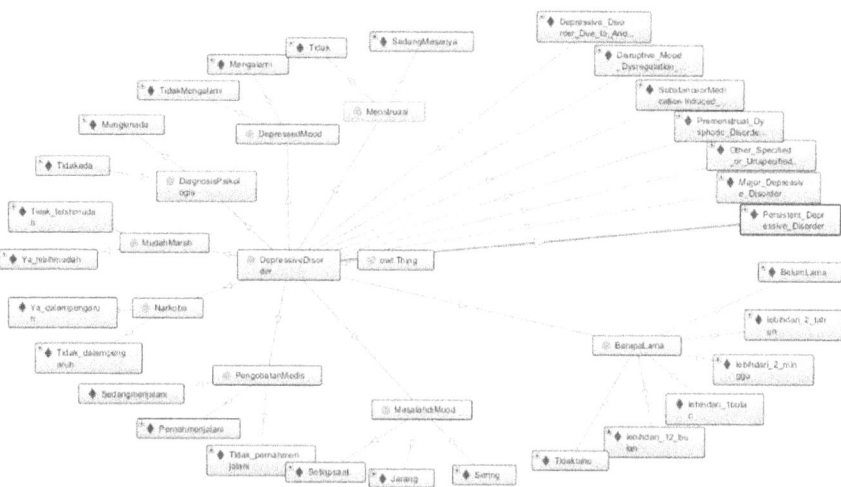

Figure 2.
Ontograph of the first scenario.

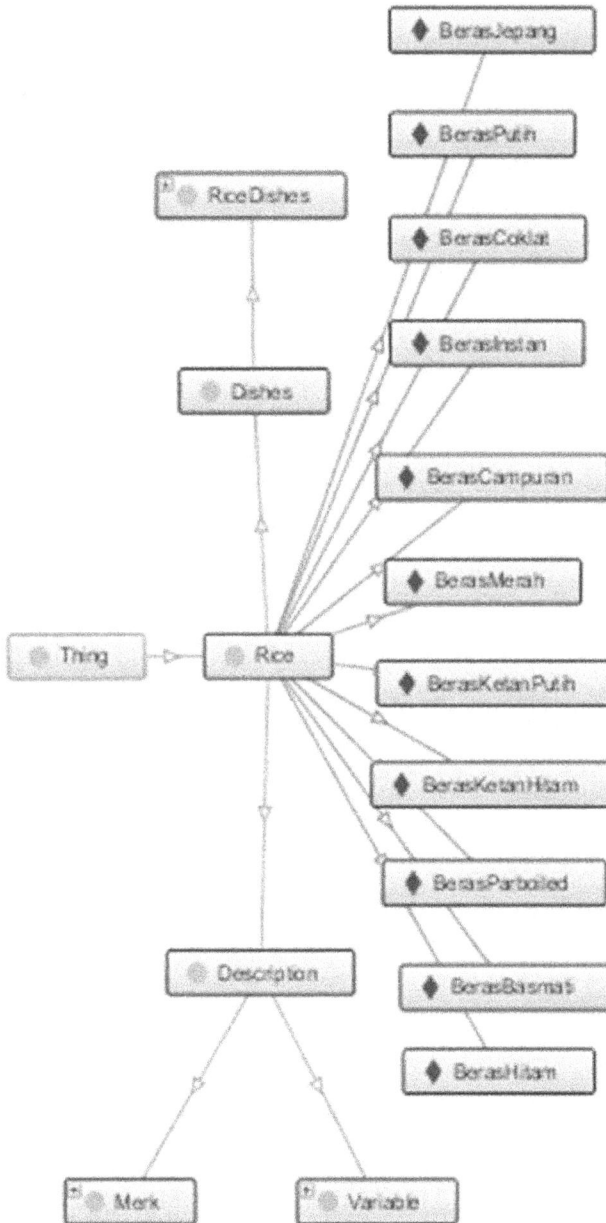

Figure 3.
Ontograph of the second scenario.

The input of the first scenario:

1. How long the user has experienced problems such as depressed mood or more irritable has six individuals, that is, more than 12 months, more than 2 weeks, do not know, more than 1 month, not long ago, and more than 2 years.

2. Depressed mood has two individuals: not experiencing and experiencing.

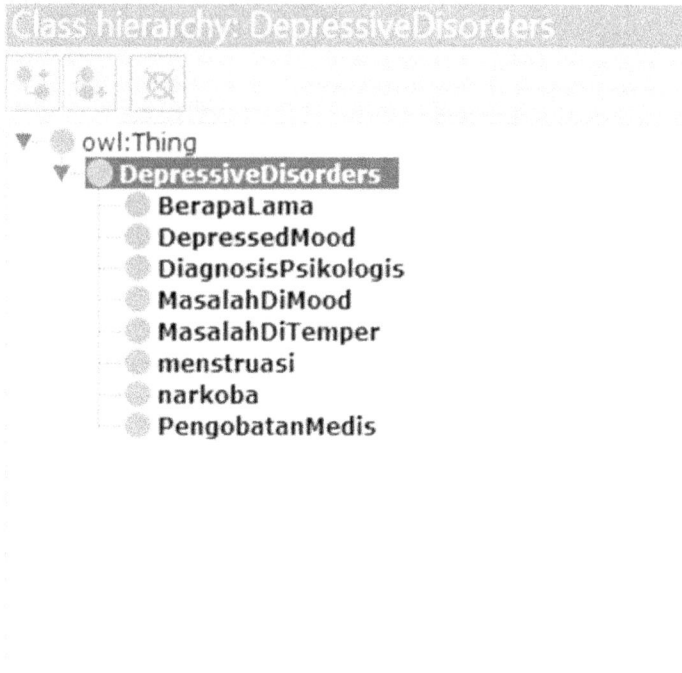

Figure 4.
Design class of the first scenario.

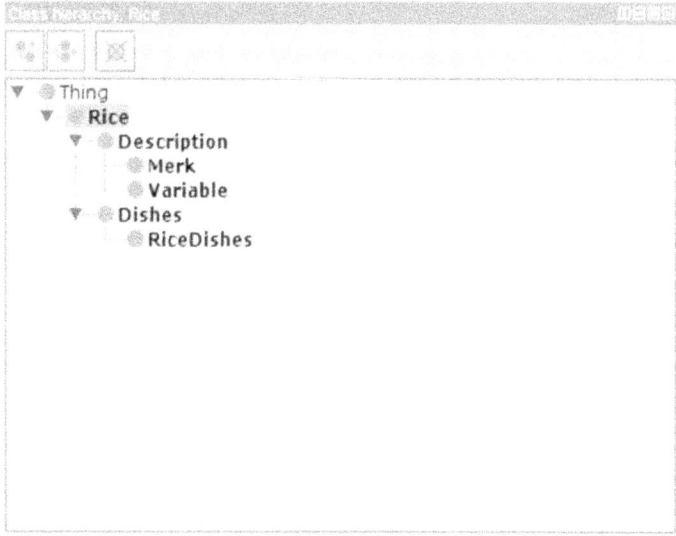

Figure 5.
Design class of the second scenario.

3. Diagnosis psychology has two individuals: none and may be present.

4. Mood problems have three individuals: all the time, rarely, and often.

5. Easy to get angry has two individuals: yes, it is easier, and no, it is not easier.

6. Menstruation has two individuals: no and moderate period. This menstruation is only experienced by women, as an initial indication of premenstrual dysphoric disorder.

7. Narcotics have two individuals: users and nonusers.

8. Medical treatment has three individuals: ever undergoing, undergoing, and never undergoing.

The output of the first scenario is the application which can find information about signs of depressive disorder. Users will choose the type of depressive disorder they want to know the information, and then the system will process. Then the desired data will appear; after getting the desired data, the user can try again to find the other data, or if there is nothing to look for, the application is complete.

The input of the second scenario is class rice has property objects and datatype properties that vary according to the characteristics of each rice. Rice class has members, namely, the type of rice. Each type of rice has different attributes. Then brands have types of rice and rice variables. Each subclass rice dish contains rice. These preparations also have object properties and datatype properties that differ according to the needs of each processed rice. Members or identifiers in the rice dish class are types of rice. Subclass brand is a subclass of the brands of rice sold in Indonesia. Each brand has an object property such as compatibility with the type of rice and the characteristic determinant of rice. Subclass variable contains variables that are used to classify the types of rice, and aroma has attributes such as the type of rice and the brand of rice that has these characteristics.

The output of the second scenario is to build an application to make it easy for users to find the most suitable type of rice so that the desired rice processing is appropriate and to find out the application of the ontology method in making expert system applications based on Android.

The next step is to convert the results from Protégé to the database. In the first scenario, we will use a CSV file where the results of Protégé are then exported to a CSV format file with entities containing individuals from depressive disorder and values of properties containing object properties. This step is shown in **Figure 6**.

Whereas in the second scenario directly using PHP, where the SPARQL query is used to retrieve data from OWL files that have been created using PHP. In this process function filters one and two function as complex character removers so that the results of the OWL can be read clearly [16–18]. This function is essential, so

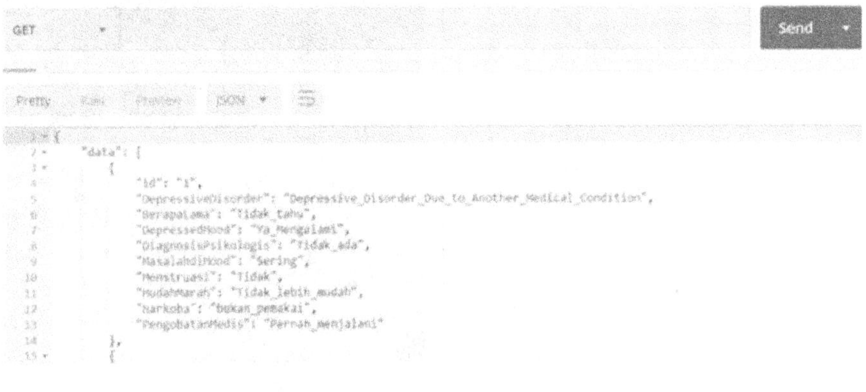

Figure 6.
SPARQL query for the first scenario.

there is no error in retrieving data from OWL. Also, this function is useful for calling data from the OWL while matching data that has been previously made according to the right results. The match function is given results for the function to

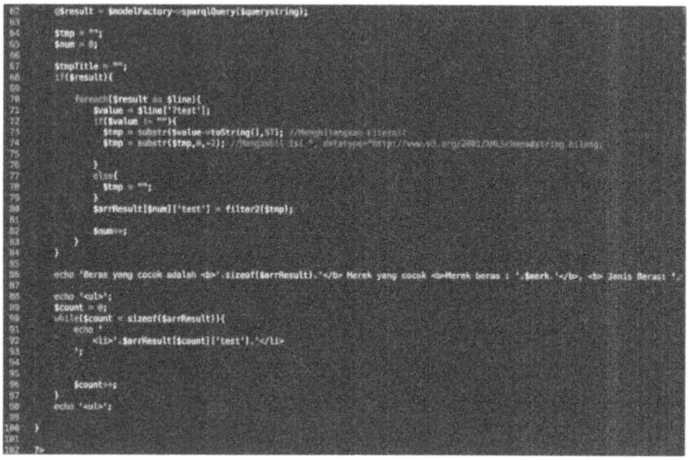

Figure 7.
SPARQL query for the second scenario.

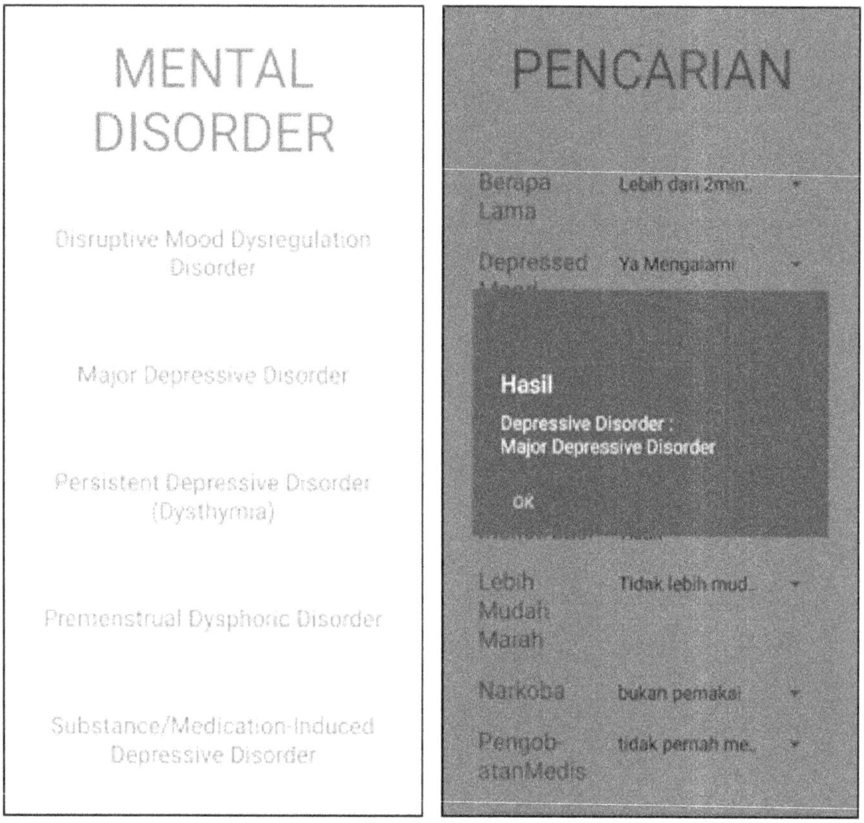

Figure 8.
Result of the first scenario.

be displayed on the application. Moreover, thus it can be concluded that the query used to retrieve data is SELECT DISTINCT * to retrieve all data using the conditions specified in the WHERE where there are conditions that must be met to choose the right results [19, 20]. This step is shown in **Figure 7**.

The following are some of the pseudocodes contained in the design:

Begin
foreach dishes
if criteria dish_rice = criteria rice
then
rice = rice_variance
end if

After doing the development and the prototype is declared complete, the implementation is done. Implementation is done when publishing the application in Play Store. After that is monitoring the users who are interested in using this application. The application created can be seen on the Google Play Store with the name MentalHelp application for the first scenario and RicheApp for the second scenario.

Figure 8 describes the Disorder search menu, wherein the user can choose the answer that is perceived by the user and then look for what kind of depressive disorder might be suitable, but the user may not find the answer sought. This image is part of the search menu where the results will appear in the selection which matches the existing data and can show the display for depressive info where there is a choice of each type of depressive disorder containing information about each type.

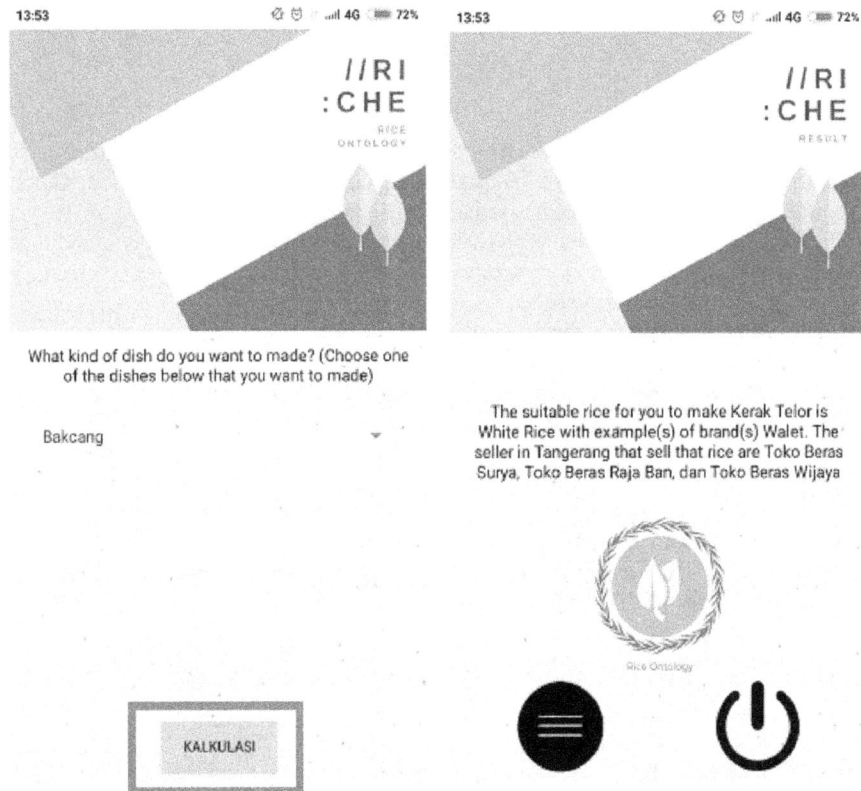

Figure 9.
Result of the second scenario.

Figure 9 is the main menu display of the application when the application starts. The main menu includes Rice Ontology, Rice Info, Rice Dishes, and Rice Seller. The Rice Selection feature makes it easy to find types of rice processed foods that are common in Indonesia. After the type of processing is selected, the user can click the "Calculate" button. After clicking the "Calculation" button, it will display the type of rice and the brand that sells the rice in Indonesia. There is a menu to return to the selection of processed types, to return to the main menu, or to exit the application.

Based on the results, the information presented in the application is complete and valid following the results of interviews from experts. In the first scenario, several developments exist after the system is created which can make it easier for people to know information about the depressive disorder. Previously to find out the type of depressive disorder, users cannot know quickly, while after this application, users can know that depressive disorder has several types. In the process of estimating the type of depressive disorder, previously it was difficult for the user to estimate whether he had a specific depressive disorder, while after this application, the user could estimate the depressive disorder that he had or whether he did not have it at all.

In the second scenario, previously in the rice selection process, the user used perceptions without specific guidelines and benchmarks, while after the application of the expert system, there could be a proper and correct reference for selecting each rice selection. In the delivery of information, the user previously conveyed information about the suitability of rice with processing based on perceptions and little knowledge of others, while after this application, information about the suitability of rice can be obtained quickly and surely whenever and wherever.

5. Conclusions

Based on the results of the research that has been carried out, these are the several conclusions for health and food problems using the ontology method, which produces the following Android-based applications:

In the first scenario: In each type of depressive disorder, it can be concluded that the characteristics of a person experiencing depressive disorder can be seen from how long someone feels it. Besides, it can be seen whether someone becomes more irritable, whether someone always feels sad, whether someone is coming for a month, whether someone has had medical treatment, whether someone is a drug user, and whether someone has another psychological diagnosis. The results of all that can be a feature of a person having a depressive disorder.

In the second scenario: Users have the convenience of being able to choose the type of rice that matches the desired rice processing and recommendations of the right type and brand through an Android-based smartphone application that can be accessed anytime and anywhere. By using this expert system, the process of selecting rice types for beginners becomes more appropriate according to expert recommendations.

Author details

Friska Natalia*, Dea Cheria and Santi Surya
Universitas Multimedia Nusantara, Tangerang, Indonesia

*Address all correspondence to: friska.natalia@umn.ac.id

IntechOpen

References

[1] Lee HK, Ko CS, Kim T, Hwang T. Fuzzy ontological knowledge system for improving RFID recognition. International Journal of Industrial Engineering. 2013;**20**(1–2):60-71

[2] Chandra D, Ferdinand FN. Ontology system design for herbal plants. International Journal of Latest Trends in Engineering and Technology (IJLTET). 2018;**10**(3):13-19

[3] Halim FN, Ferdinand CSK. Ontology-based decision support system for hypersensitivity disorder allergy. ICIC Express Letters. 2018;**12**(8): 847-854

[4] El-sappagh S, Elmogy M. Fuzzy ontology modeling for case base knowledge in diabetes mellitus domain. Engineering Science and Technology an International Journal. 2017;**20**(3): 1025-1040

[5] Rajendran VV, Swamynathan S. MOSS-IR: Multi-ontology based search system for information retrieval in e-health domain. Procedia Computer Science. 2015;**47**:179-187

[6] Helmy T, Al-Nazer A, Al-Bukhitan S, Iqbal A. Health, food and user's role ontologies for personalized information retrieval. Procedia Computer Science. 2015;**52**:1071-1076

[7] Dennis A. System Analysis and Design. 6th ed. America: Wiley Inc; 2014

[8] O'Brien, James A. dan George M. Marakas. Management Information Systems, 10th ed. McGraw-Hill/Irwin, New York, 2011

[9] Kelly Rainer R. Brad Prince. Introduction to Information System. 7th ed. America: Wiley Inc; 2017

[10] Association AP. Diagnostic and Statistical Manual of Mental Disorders, Fifth Edition (DSM-5). America: American Psychiatric Association; 2013

[11] Horridge MA. Practical guide to building OWL ontologies using protege 4 and CO-ODE tools Edition 1.3. 2011

[12] Lee HK, Ferdinand FN, Kim T. Fuzzy ontology-based supply partner matching. ICIC Express Letters: International Journal of Research and Surveys. 2011;**5**(9B):3329-3334

[13] Xiolong L. Software Engineering and Information Technology. World Scientific; 2015

[14] Upward A, Jones PH. An ontology for strongly sustainable business models: Defining an enterprise framework compatible with natural and social science. Organization & Environment, Special Issue: Business Models for Sustainability: Entrepreneurship, Innovation, and Transformation (On-Line First). 2015:1-27

[15] Neji H, Bouallegue R. Ontology for mobile phone operating. 2012

[16] Workman M. Semantic Web: Implications for Technologies and Business Practices 1st ed. Springer; 2016

[17] Lee B, James H, Lassila O. The semantic web. Scientic American. 2001: 29-37

[18] Pollock TJ. Semantic Web for Dummies. Indiana: Wiley Publishing Inc.; 2009

[19] Connoly T, Begg C. Database System: A Practical Approach to Design, Implementation, and Management. America: Publisher Pearson; 2010

[20] Elve AT, Preisig HA. From ontology to executable program code. Computers and Chemical Engineering. 2019;**122**: 383-394

Taxonomy and Ontology Management Tools: A General Explanation

Sukumar Mandal

Abstract

The World Wide Web is the result of a radical new way of thinking about sharing information. This idea seems familiar now, as the web itself has become pervasive. But this radical new way of thinking has even more profound ramifications when it is applied to a web of data like the semantic web. These ramifications have driven many of the design decisions for the semantic web standards and have a strong influence on the craft of producing quality semantic web applications. Until several years ago, the semantic web was primarily in a research phase. The new implementations of it were mainly to demonstrate the potential of the idea. While much of the activity related to semantic technology still takes place within the academic community, there are now real-world examples of the technology to use as a model. This paper has selected the matured level open-source tools for management of taxonomy and ontology in semantic web environment. Apart from this it also presents some important snapshot available in an online environment for designing and developing the taxonomy and ontology. This is very helpful to the users in using the concept of science and technology in cloud computing and online and offline environment.

Keywords: ontology, semantic web, TemaTres, visual vocabulary, and taxonomy

1. Introduction

This is the age of technology where digital information resources are increasing. It is a good concept for creating both the present and next-level automated and digital library system because it can highly be performed in the semantic web environment for executing the functional activities of libraries and any information resource centers. It helps the researchers for reviewing their literature. Here users can easily access and understand the terms and their relation from these interfaces for the better management of library and information services. It manages the big data in semantic web-linked data environment. It is also possible to manage the bibliographic link data in the visual vocabulary format, so the users can easily download the content as well as information. This is a web-based architecture in the Internet. Semantic web is one of the important aspects in cloud computing [1]. A lot of linking of web resources are available in an online environment [2]. The idea of a web of information was once a technical idea accessible only to a highly trained elite of information professionals: IT administrators, librarians, information architects, and other resources [3]. It is known as big data concept in semantic web level. It is based on open-source standards and formats including SPARQL, RDF, JSON, HTML,

XHTML, DTD, METS, MODS, etc. for managing the different web resources in any institutions or in any library [4]. These formats help to draw the visual graph of each uniform resource identifier for maintaining and managing the big data in the Internet [5]. Visual vocabulary is also an important concept in semantic web because it stores the large number of data and their linking also [6]. It is possible to manage the relations of different subjects by their specific identities, properties, and entities [7]. Big data is easily managed by using the open-source software and open standard formats in an online environment [8]. Users can easily access their necessary information by using this semantic web software in an Internet environment because it can manage the controlled vocabularies and their relationships of each element of any subject fields [9]. So, obviously it can save the time of reader and library professionals also [10]. It can manage the N-Triples, a format for storing and transmitting data, and Terse RDF Triple Language (Turtle), and XML provides an elemental syntax for content structure within documents yet associates no semantics with the meaning of the content contained within [11]. XML is not at present a necessary component of semantic web technologies in most cases, as alternative syntaxes exist, such as Turtle [12]. The World Wide Web contains many billions of pages. The SNOMED CT medical terminology ontology alone contains 370,000 class names, and existing technology has not yet been able to eliminate all semantically duplicated terms [13]. Any automated reasoning system will have to deal with truly huge inputs [14]. The objective of this chapter is very simple to highlight some popular taxonomy, ontology, and visual vocabulary through open-source tools for the users as well as library professionals. The important tools have been discussed in the next section. It also shows how it works in the web environment.

2. Objectives

The essential objectives of this study are explained as below:

i. To explore the modern tools and techniques those are available in an online digital environment for easy generation of taxonomy and ontology in different aspects.

ii. To highlights and visualize some terms by using these open source tool.

iii. To show the taxonomy concepts in an academic and research environment of different relations both technically and graphically.

iv. To design an integrated framework in an offline and online environment by using the open source tool and technique for each an every concept and their relations for the management of information resources.

3. Methodology

The process and methods of this study are one of the important aspects for designing and creating taxonomy and ontology. In this original research paper, the matured level open-source tools and techniques have been selected for the creation of visual vocabulary. Apart from this it also explores how to construct a visual thesaurus in an offline and online environment on Ubuntu operating system for easy constructing of the different terms and relations such as narrower term, broader term, related term, used for, scope note, and so on. So, this paper consists of two parts such as online and offline for term relations.

4. Visuwords

Visuwords is an online visual graphical dictionary as well as controlled vocabulary tool. Actually it provides the visual interface of each word and its respective meaning on the basis of English grammar. This is very user-friendly for definition of any words from any subjects can be represented in a separate color dashboard. Wheel and zoom are possible by using the mouse. It explores the word synonym, derivation, and antonym to extend the meaning of any facets available in an web-based environment (https://visuwords.com/library). **Figure 1** represents the Visuwords interface in an online environment for visual graphic dictionary. It is possible to easily identify the noun, verb, adjective, and adverb for construction the thesaurus against in a word.

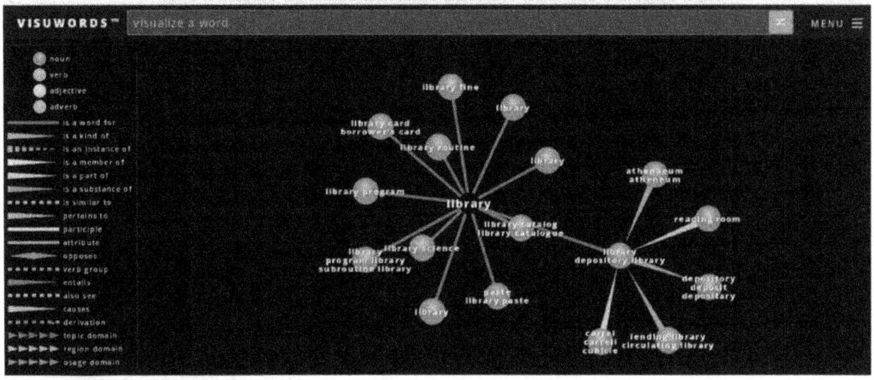

Figure 1.
Visuwords interface for visual graphic dictionary (https://visuwords.com/library).

5. Linked open vocabularies

Linked Open Vocabularies is a new concept in semantic web ontology. It consists of large amount of dataset around 660 vocabularies as on March 10, 2019. It fully supports the OWL and RDF framework for better management and identification of the right link available in an online database. SPARQL endpoint is easily accessed by using this interface. Properties and classes can be done on the basis of ontology concept that is subject, predicate, and object (https://lov.linkeddata.es/dataset/lov/). Here all the dataset belongs to the vocabulary content or element type. **Figure 2** represents the Linked Open Vocabularies interface in semantic web for different elements and their relationship.

6. WebVOWL

WebVOWL is a web-based ontology visualization open-source tool and is very interactive. This is written by high-level programming language named as Java which implements the visual notation and graphical interface of different nodal points in web ontology language. It explores the creation of ontology and combined forced graph in different formats such as JSON, RDF, and XML and other relevant file formats in an integrated semantic web environment (http://visualdataweb.de/webvowl/). **Figure 3** represents the WebVOWL interface in semantic web for linking a custom ontology based on URL and URI.

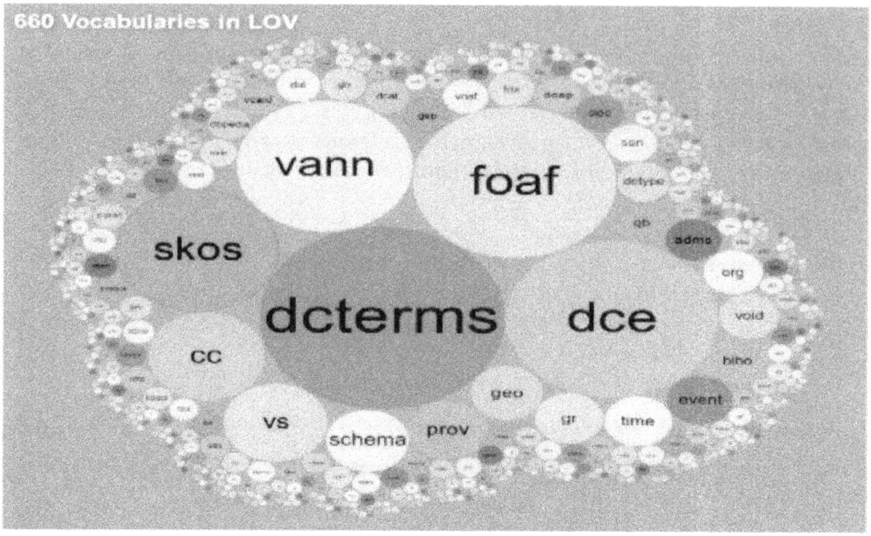

Figure 2.
Linked open vocabularies interface in semantic web ontology (https://lov.linkeddata.es/dataset/lov/).

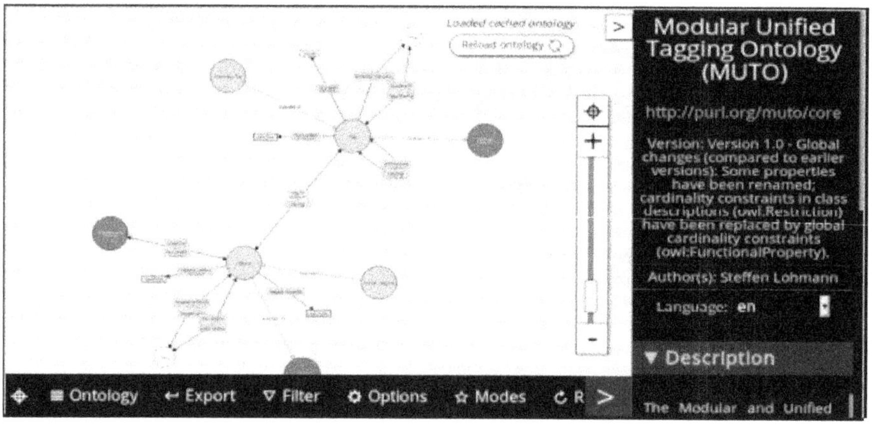

Figure 3.
WebVOWL interface in semantic environment (http://visualdataweb.de/webvowl/).

7. NavigOWL

NavigOWL is a graph visualization open-source Java-based tool in web ontology environment. This tool is a very interactive high-performance graph layout in semantic web environment. Here all ontologies are structured based which facilitate the patrons in thinking the mental map in an ontology environment. Protégé can be graphically represented by using this plug-in for better management of link and their elements (http://home.deib.polimi.it/hussain/navigowl/index. html). **Figure 4** represents the NavigOWL interface in web ontology environment. It loads and uploads the RDF/OWL ontology file for creation of graph and taxonomy in different nodes and edges.

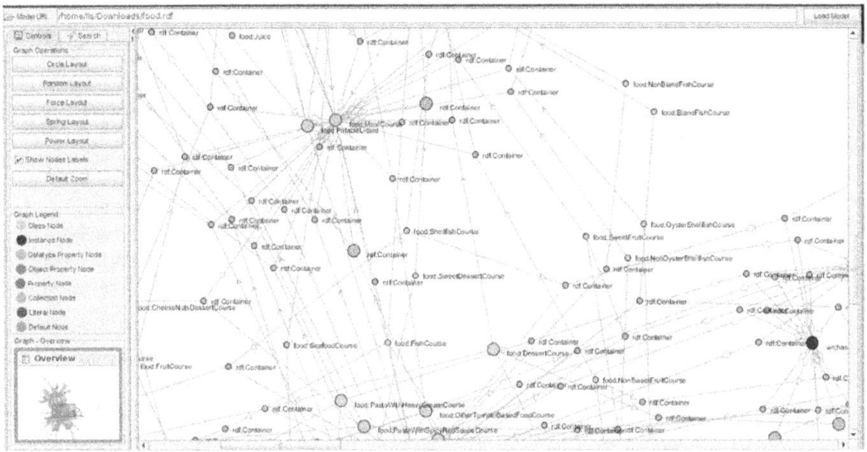

Figure 4.
NavigOWL interface in web ontology.

8. Unilexicon vocabulary

Unilexicon is a visual online controlled vocabulary and thesaurus construction server. This is a high-level open-source software for designing taxonomy and classification. It fully supports the faceted navigation and tagging for document and knowledge classification on the basis of simple knowledge organization system (SKOS). Institutional digital repository can easily be designed and developed by using the metadata and semantic search tool. This is also known as record management tool (https://unilexicon.com/). **Figure 5** represents the Unilexicon vocabulary interface. This is independent of content management system by using tag and facet count concept.

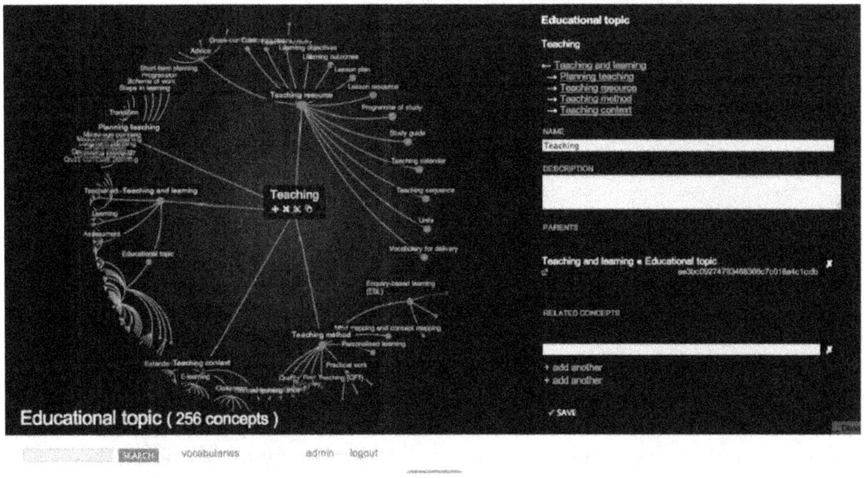

Figure 5.
Unilexicon vocabulary interface in taxonomy (https://unilexicon.com/).

9. Visual vocabulary TemaTres

TemaTres is a web-enabled open-source thesaurus construction software and written in PHP programming language. It is also known as vocabulary server for management formal representations of knowledge, thesauri, taxonomies, and multilingual vocabularies [15]. It has many features including SPARQL Protocol and RDF Query Language, Meta-terms (define facets, collections or arrays of terms), support for multilingual thesaurus, expose vocabularies with powerful web services, search terms suggestion, display terms in multiple deep levels in the same screen, search expansion, vocabulary harmonization, relationship between terms (BT/NT, USE/UF, RT), no limits to number of terms, alternative labels, levels of hierarchy, etc., systematic or alphabetical navigation, complete export in XML format (Zthes, TopicMaps, MADS, Dublin Core,VDEX, BS 8723, SiteMap, SQL), complete export in RDF format, complete export in txt, scope notes, historical and bibliographical notes, user management, terms and user supervision, duplicates terms control, free terms control, quality assurance functions (Duplicates and free terms, ilegal relations), multilingual interface, easy install, utility to import thesauri from tabulated textfiles, unique code for each term, terminology mapping with multilingual, term reports for editors, workflow like candidate, accepted and rejected terms, allow to create user-defined relationships, allow to define published and hidden labels, relationships between terms and web entities, export to WXP (WordPress XML) and import and export data in Skos-core (**Figure 6**).

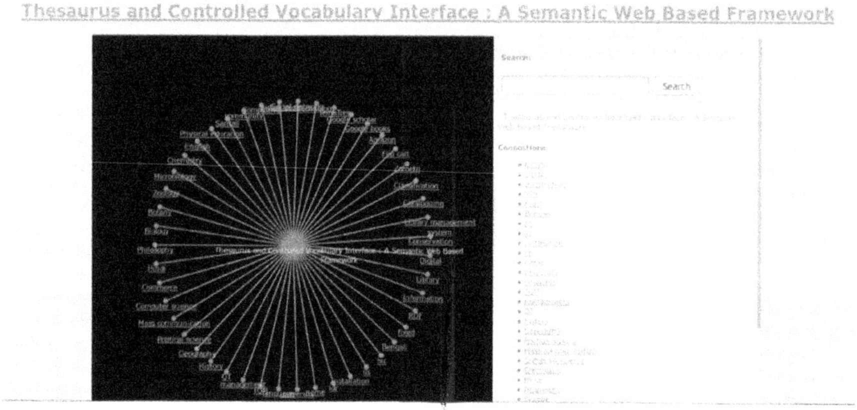

Figure 6.
Visual vocabulary interface in TemaTres.

10. Conclusion

Vocabulary creation is important for the users in any library including academic, public, and research libraries. It can manage and maintain the narrower terms and broader terms in respect to one term of a specific subject like philosophy, Bengali, history, chemistry, etc. Search interfaces of visual vocabulary are very sophisticated, and here quick search facilities are available also. The alphabetical index part contains each and every term, including synonyms, quasi-synonyms, and antonyms, occurring in the systematic part, along with its narrower terms. It controls the different terms used in indexing, providing a means of translating the natural language authors, indexers, and enquirers into a more constrained language used for indexing and retrieval. Visual vocabulary is one of the important concepts of formulation of different words and their associated terms. All the terms and correlated terms against one specific subject are to be appeared in a visual vocabulary interface. It can manage the different terms including broader terms, narrower terms, and both preferred and non-preferred terms. In this section overview, the concept of a visual vocabulary is a strategy that draws inspiration from the text retrieval community and enables efficient indexing for different terms of a specific item types. Since the occurrence of a given word tends to be sparse across different documents, an index that maps words to the files in which they occur can take a keyword query and immediately produce relevant content. Obviously, it can conclude that the above tools highly overview the taxonomy and ontology in semantic web environment for easy access and download of metadata as well as full-text resources and their measures by graph. Thesaurus, taxonomy, and ontology are properly managed by using these tools for better management and retrieval of linked and resources in the web environment.

Author details

Sukumar Mandal
Department of Library and Information Science, The University of Burdwan, Golapbag, West Bengal, India

*Address all correspondence to: sukumar.mandal5@gmail.com

IntechOpen

References

[1] Alfaries A, Bell D, Lycett M. Motivating service re-use with a web service ontology learning. International Journal of Web Information Systems. 2013;**9**(3):219-241. DOI: 10.1108/IJWIS-12-2012-0035

[2] Benslimane SM, Malki M, Bouchiha D. Maintaining web application: An ontology-based reverse engineering approach. International Journal of Web Information Systems. 2009;**5**(4):495-517. DOI: 10.1108/17440080911006225

[3] Bygstad B, Ghinea G, Klæboe G-T. Organisational challenges of the semantic web in digital libraries: A Norwegian case study. Online Information Review. 2009;**33**(5):973-985. DOI: 10.1108/14684520911001945

[4] Calaresu M, Shiri A. Understanding semantic web: A conceptual model. Library Review. 2015;**64**(1/2):82-100. DOI: 10.1108/LR-09-2014-0097

[5] Chi Y-L, Chen H-C. Ontology and semantic rules in document dispatching. The Electronic Library. 2009;**27**(4):694-707. DOI: 10.1108/02640470910979633

[6] Esserhrouchni OEI, Frikh B, Ouhbi B, Ibrahim IK. Learning domain taxonomies: The TaxoLine approach. International Journal of Web Information Systems. 2017;**13**(3):281-301. DOI: 10.1108/IJWIS-04-2017-0024

[7] Khan SA, Bhatti R. Semantic web and ontology-based applications for digital libraries: An investigation from LIS professionals in Pakistan. The Electronic Library. 2018;**36**(5):826-841. DOI: 10.1108/EL-08-2017-0168

[8] Ko YM, Song MS, Lee SJ. Construction of the structural definition-based terminology ontology system and semantic search evaluation. Library Hi Tech. 2016;**34**(4):705-732. DOI: 10.1108/LHT-08-2016-0090

[9] Lausen H, Ding Y, Stollberg M, Fensel D, Hernández RL, Han S-K. Semantic web portals: State-of-the-art survey. Journal of Knowledge Management. 2005;**9**(5):40-49. DOI: 10.1108/13673270510622447

[10] Llanes-Padrón D, Pastor-Sánchez J-A. Records in contexts: The road of archives to semantic interoperability. Program. 2017;**51**(4):387-405. DOI: 10.1108/PROG-03-2017-0021

[11] Navarro-Galindo JL, Samos J. The FLERSA tool: Adding semantics to a web content management system. International Journal of Web Information Systems. 2012;**8**(1):73-126. DOI: 10.1108/17440081211222609

[12] Nguyen H-M, Nguyen H-Q, Tran K-N, Vo X-V. GeTFIRST: Ontology-based keyword search towards semantic disambiguation. International Journal of Web Information Systems. 2015;**11**(4):442-467. DOI: 10.1108/IJWIS-06-2015-0019

[13] Nguyen Q-M, Cao T-D. A novel approach for automatic extraction of semantic data about football transfer in sport news. International Journal of Pervasive Computing and Communications. 2015;**11**(2):233-252. DOI: 10.1108/IJPCC-03-2015-0018

[14] Salvadori IL, Huf A, Oliveira BCN, Mello RS, Siqueira F. Improving entity linking with ontology alignment for semantic microservices composition. International Journal of Web Information Systems. 2017;**13**(3): 302-323. DOI: 10.1108/IJWIS-04-2017-0029

[15] Mandal S. Developing thesaurus construction through Tematres for the college libraries under the University of Burdwan. International Journal of English Language, Literature in Humanities. 2016;**4**(6):302-316

www.ingramcontent.com/pod-product-compliance
Lightning Source LLC
Chambersburg PA
CBHW081240190326
41458CB00016B/5862